Collins

NEW COMPLETE WIRING & LIGHTING

Collins

NEW COMPLETE WIRING & LIGHTING

Albert Jackson & David Day

This edition published in 2006 by Collins,
an imprint of HarperCollins Publishers
77–78 Fulham Palace Road
Hammersmith
London W6 8JB

The Collins website address is www.collins.co.uk

12 11 10 09 08 07 06
7 6 5 4 3 2 1

A catalogue record for this book is available from
the British Library

ISBN-13 978 0 00 723193 5
ISBN-10 0 00 723193 8

Colour origination by Colourscan, Singapore
Printed and bound by CIP Bath

The material for this book was created originally for
HarperCollins Publishers Ltd by Jackson Day Jennings Ltd
trading as Inklink. This edition was created by Jackson Day.
Most of the text and illustrations in this book also appear
in the *Collins Complete DIY Manual*.

Authors Albert Jackson and David Day

Design Simon Jennings and Elizabeth Standley

Text editor Peter Leek

Consultant John Dees

Illustrations editor David Day

Illustrators Robin Harris and David Day

Additional illustrations Brian Craker, Michael Parr and Brian Sayers

Photographers Ben Jennings and Colin Bowling

Additional photography Neil Waving and Paul Chave

Proofreader Mary Morton

Research editor and technical consultant Simon Gilham

PLEASE NOTE
Great care has been taken to ensure that the information contained in this book is accurate. However, the law concerning Building Regulations, local bylaws and related matters is neither static nor simple. A book of this nature cannot replace specialist advice in appropriate cases and therefore no responsibility can be accepted by the publisher or by the authors for any loss or damage caused by reliance upon the accuracy of such information.

CONTENTS

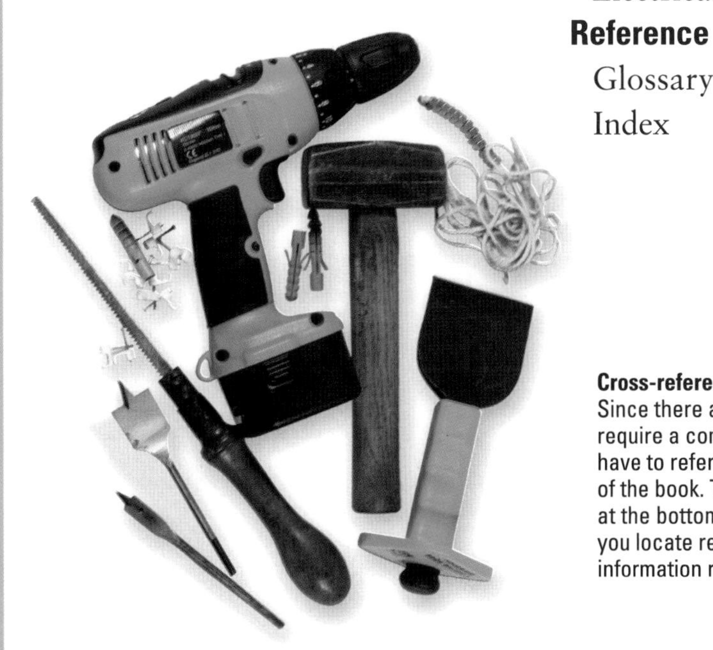

Cross-references
Since there are few projects that do not require a combination of skills, you will have to refer to more than one section of the book. The list of cross-references at the bottom of each page will help you locate relevant sections or specific information related to the job in hand.

Reducing electricity bills

Pressures from all sides urge us to conserve energy – and this applies just as much to electricity as to fossil fuels such as coal, oil and gas. But even without such encouragement, our quarterly electricity bills would provide stimulus enough to make us find ways of using less power.

Nobody wants to live in a poorly heated or dismally lit house, without the comforts of hot water, refrigeration, television and other conveniences – but it is often possible to identify where energy is wasted and then find ways to reduce waste without compromising your comfort or pleasure.

Avoid false economy

Whether you do your own wiring or employ a professional, don't attempt to economize by installing fewer sockets than you really need. When you rewire a room, fit as many as you may possibly use. The inconvenience and expense later on of running extra cable and disturbing decoration will far outweigh the cost of an extra socket or two.

Similarly, don't restrict your use of lighting unnecessarily. Used sensibly, lights consume relatively little power, so it isn't worth risking accidents – for example, on badly lit stairs. Nor need you strain your eyes in the glare from a single light hanging from the ceiling, when extra lighting could provide comfortable and attractive background illumination.

Fitting controls to save money

As the chart opposite clearly shows, heating is by far the biggest consumer of domestic power. One way to reduce your electricity bills is to fit devices that regulate the heating in your home to suit your lifestyle, maintaining comfortable but economic temperatures.

Thermostats

Most modern heating has some form of thermostatic control – a device that will switch power off when surroundings reach a certain temperature. Many thermostats are marked out simply to increase or decrease the temperature, in which case you have to experiment with various settings to find the one that suits you best. If the thermostat settings are more precise, try 18°C (65°F) for everyday use – although elderly people are more comfortable at about 21°C (70°F).

As well as saving you money, an immersion-heater thermostat prevents your water from becoming dangerously hot. Set it at 60°C (140°F). See far right for Economy 7 setting.

Time switches

Even when it's thermostatically controlled, heating is expensive if run continuously – but you can install an automatic time switch to turn it on and off at preset times, so you get up in the morning and arrive home in the evening to a warm house. Set it to turn off the heating about half an hour before you leave home or go to bed, as the house will take time to cool down.

A similar device will ensure that your water is at its hottest when needed.

Monitoring consumption

Keep an accurate record of your energy saving by taking weekly readings. Note the dates of any measures taken to cut power consumption, and compare the corresponding drop in meter readings.

Digital meters

Modern meters display a row of figures or digits that represent the total number of units consumed since the meter was installed. To calculate the number of units used since your last electricity bill, simply subtract the 'present reading' shown on your bill from the number of units now shown on the meter. Make sure that the bill gives an actual reading and not an estimate (which is indicated by the letter 'E' before the reading).

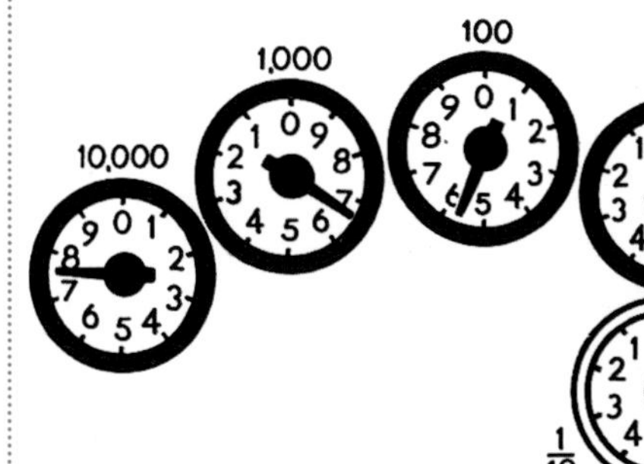

Reading a dial meter
Write down your reading from right to left. This meter records 76,579 units.

• Shopping around
Getting quotes from various electricity companies may help you find a better deal. Tell each company how many units you use on average each year, and see if you can make a saving by changing your supplier. Consult the following web sites:
www.uswitch.com
www.ukpower.co.uk

• Insulation
Measures taken to save energy will have little effect unless you insulate your house as well as the hot-water cylinder and pipework. You can do most of the work yourself for a relatively modest outlay and a little effort.

OFF-PEAK RATES

Electricity is normally sold at a general-purpose rate, every unit used costing the same; but if you warm your home with storage heaters and heat your water electrically, then you can take advantage of the economical off-peak tariff. This system, called Economy 7, allows you to charge storage heaters and heat water at less than half the general-purpose rate for seven hours, starting between midnight and 1 a.m. Other appliances used during that time get cheap power too, so more savings can be made by running the dishwasher or washing machine after you've gone to bed. Each appliance must, of course, be fitted with a timer. The Economy 7 daytime rate is higher than the general-purpose one, but the cost of running 24-hour appliances such as freezers and refrigerators is balanced since they also use cheap power for seven hours.

For full benefit from off-peak water heating use a cylinder that holds 182 to 227 litres (40 to 50 gallons), to store as much cheap hot water as possible. You will need a twin-element heater or two separate units. One heater, near the base of the cylinder, heats the whole tank on cheap power; another, about half way up, tops up the hot water during the day. Set the night-time heater at 75°C (167°F), the daytime one at 60°C (140°F).

The electricity companies provide Economy 7 customers with a special meter to record daytime and night-time consumption separately, plus a timer that automatically switches the supply from one rate to the other.

Reading dial meters

The principle of a dial meter is simple. Ignore the dial marked 1/10, which is only for testing. Start with the dial indicating single units (kWh) and, working from right to left, record the readings from the 10, 100, 1000 and finally 10,000 unit dials. Note the digits the pointers have passed. If a pointer is, say, between 5 and 6, record 5. If it's right on a number, say 8, check the next dial on the right: if that pointer is between 9 and 0, record 7; if it's past 0, record 8. Also, remember that adjacent dials revolve in opposite directions, alternating along the row.

SEE ALSO: Immersion heaters 38–9, Storage heaters 40–1

Running costs of your appliances

Apart from the standing charge and any hire-purchase payments, your electricity bill is based on the number of units of electricity you have consumed during a given period. Each unit represents the amount used in one hour by a 1kW appliance. An appliance rated at 3kW will use the same amount of energy in 20 minutes.

TYPICAL RUNNING COSTS

Appliance	Typical usage	No. of units	Appliance	Typical usage	No. of units
Cooker	Cooks 1 day's meal for four people.	2½	Iron	In use for 2 hours.	1
Microwave	Cooks 2 joints of meat.	1	Vacuum cleaner	Works for 1½–2 hours.	1
Slow cooker	Cooks for 8 hours.	1	Cooker hood	Runs for 24 hours continuously.	2
Storage heater (2kW)	Provides 1 day's heating.	11	Extractor fan	Runs for 24 hours continuously.	1
Fan heater or bar fire (2kW)	Provides heat for 1 hour.	2	Hairdryer	Runs for 2 hours.	1
Immersion heater	Supplies 1 day's hot water for a family of four.	9	Shaver	Gives 1800 shaves.	1
Instant water heater	Heats 2 to 3 bowls of washing-up water.	1	Single overblanket	Warms the bed for 1 week.	2
Instant shower	Gives 1 to 2 showers.	1	Single underblanket	Warms the bed for 1 week.	1
Dishwasher	Washes 1 full load.	2	Power drill	Works for 4 hours.	1
Automatic washing machine	Washes 1 full load with prewash.	2½	Hedge trimmer	Trims for 2½ hours.	1
Tumble dryer	Dries 1 full load.	2½	Cylinder lawn mower	Cuts grass for 3 hours.	1
4cu ft refrigerator	Keeps food fresh for 1 week.	7	Hover mower	Cuts grass for 1 hour.	1
6cu ft freezer	Maintains required temperature for 1 week.	9	Stereo system	Plays for 8 hours.	1
Heated towel rail	Warms continuously for 4 hours.	1	Colour TV	Provides 6 hours' viewing.	1
Electric kettle	Boils 40 cups of tea.	1	VCR	Records for 10 hours.	1
Coffee percolator	Makes 75 cups of coffee.	1	100W bulb	Gives 10 hours' illumination.	1
Toaster	Toasts 70 slices of bread.	1	40W fluorescent strip light	Provides 20 hours' illumination.	1

● **Typical running costs**
The table gives estimates of units used by typical appliances in normal use. These figures may be affected by variables such as room temperature. Also, figures for certain appliances, such as LCD televisions, may be lower than those shown here. However, the table will help you determine which appliances are likely to add most to the cost of electricity in your home.

● **Green efficiency labels**
When you're shopping for new appliances, take advantage of the labelling system, which includes guidance on energy efficiency. The choice of an 'A' rating can make considerable savings over the life of the appliance.

SEE ALSO: Heaters 35, Towel rail 35, Cooker hoods 36, Extractor fans 36, Kitchen appliances 36, Water heater 36, Cookers 37–8, Immersion heaters 38–9, Storage heaters 40–1, Shower 47, Lighting 47–57

Understanding the basics

Many people imagine the electrical circuits of a house are complicated – but the circuitry is, in fact, based on simple principles.

For any electrical appliance to work, the power must have a complete circuit – the electricity must be able to flow along a wire from its source (a battery, for instance) to the appliance (say a light bulb) and then back to the source along another wire. If the circuit is broken at any point, the appliance will stop working – the bulb will go out. Breaking the circuit – and restoring it as required – is what a switch is for. When the switch is in the 'on' position, the circuit is complete and the bulb or other appliance operates. Turning the switch off makes a gap in the circuit, so the electricity stops flowing.

Although mains electricity is much more powerful than that produced by a battery, it operates in exactly the same way, flowing through live or 'phase' wires linked to every light, socket outlet and fixed electrical appliance in your home. The current flows back out of the building through the neutral wires.

A basic circuit
Electricity runs from the source (battery) to the appliance (bulb) and then returns to the source. A switch breaks the circuit to interrupt the flow of electricity.

Earthing

Any material through which electricity can flow is known as a conductor. Most metals conduct electricity well, which is why copper is used for electrical wiring.

However, the earth itself – the ground on which we stand – is also an extremely good conductor, which is why electricity always flows into the earth whenever it has an opportunity to do so, taking the shortest available route. This means that if you were to touch a live conductor, the current would divert and take the short route through your body to the earth.

A similar thing can happen if a live wire accidentally comes into contact with any exposed metal component of an appliance, including its casing. To prevent this, a third wire is included in the wiring system and connected to the earth, usually via the outer casing of the electricity company's main service cable. This third wire – called the earth wire – is attached to the metal casing of some appliances and to special earth terminals in others, providing a direct route to the ground should a fault occur. This sudden change of route by the electricity – known as an earth fault – causes a fuse to blow or circuit breaker to operate, cutting off the current.

Double insulation
Appliances that are double-insulated – which usually means they have a plastic casing that insulates the user from metal parts that could become live – must not be earthed with a third wire. A square within a square, either printed or moulded on an appliance, means it is double-insulated and its flex does not need an earth wire.

Building Regulations on electrical wiring

Regulations – known as Part P of the Building Regulations – have been introduced to promote better standards. These regulations do not prevent the DIY worker from undertaking electrical wiring, but they put strict limits on what can be done without supervision and inspection.

All local authorities have Building Control Officers (BCOs) who are responsible for monitoring the regulations. You should therefore contact your BCO to ascertain how exactly your particular authority applies the regulations.

Certain tasks can be undertaken without having to notify the authority, and most BCOs do not require any communication or paperwork for 'non-notifiable' electrical work. However, where similar work is carried out in locations such as kitchens and bathrooms or the wiring will be involved or extensive, then the work becomes 'notifiable' and must be discussed with the BCO before it is carried out.

Non-notifiable work

A DIY worker can do the following, anywhere in the home without having to inform the BCO in advance:

- Replace sockets, fused connection units, switches and ceiling roses.
- Replace damaged cable for a single circuit.
- Refix or replace enclosures (mounting boxes for sockets and switches) on existing circuits.
- Provide mechanical protection in the form of conduit and plastic channel.

Also without having to inform the BCO in advance, a DIY worker can do the following anywhere in the home except in kitchens, bathrooms and utility rooms and in special locations – such as rooms with a bathtub or shower, swimming pools and paddling pools, hot-air saunas and outdoors – or when installing extra-low-voltage lighting:

- Add new light fittings and switches to existing circuits.
- Add new socket outlets to existing ring circuits and radial circuits.
- Add fused spurs to existing ring and radial circuits.

Notifiable work

The BCO must be notified before a DIY worker undertakes any electrical work not listed above or if the work is categorized as one of the exceptions described above.

Notifiable work
To help you decide which electrical work you can do yourself, all notifiable work shown in this book is marked with this symbol.

COMPLYING WITH THE REGULATIONS

It is not necessary to involve the BCO when any work is undertaken by a professional electrician – that is a competent person registered with an electrical self-certification scheme. This person will deal with all the paperwork required by the authority and, on completion of the work, should give you a signed Building Regulations Self-certification Certificate and a completed Electrical Installation Certificate.

If you feel competent to do any notifiable work yourself, you must tell your local BCO in advance exactly what you propose to do. Having obtained permission to proceed, once you have completed the work you must ask the BCO to send an inspector to test the installation and issue a certificate. A fee will be charged for inspection and testing. Most BCOs will offer some concession – such as including the electrical inspection in the general inspection costs for building a new extension. Some BCOs may ask you to arrange for a competent electrician to inspect and test the work.

Procedures are laid down for appeals, determinations, relaxations and dispensations, but the common-sense approach is to accept any advice or instruction given by your Building Control Officer.

Provided you are competent to undertake non-notifiable work, you may proceed without supervision so long as the methods used comply with the the IEE Wiring Regulations (BS7671). Good workmanship and the use of proper materials are fundamental to these regulations. You should also keep a record of your work in the form of a Minor Electrical Works Certificate, which you can pass on to interested parties should you decide to sell your home in the future. Selling a house without appropriate paperwork may introduce delays and difficulties.

The methods and materials suggested in this book comply with the Wiring Regulations, but if you have any doubts about your ability to satisfy their requirements, then use this book to study the work involved so you can brief a professional electrician and agree an appropriate price for the job. Ensure that any electrician you hire is a member of an authorized competent-person self-certification scheme.

SEE ALSO: Earth bonding 10, 17, Flex 12, Fuses 19, Circuits 21, Cables 22, Running cable 23–5, Final tests 67

With safety in mind

Throughout this chapter you will find frequent references to the need for safety while working on your electrical system – but it cannot be stressed too strongly that you must also take steps to safeguard yourself and others who will later be using the system. Faulty wiring and appliances are dangerous, and can be lethal. Whenever you are dealing with electricity, the rule must be 'safety first'.

- Never inspect or work on any part of an electrical installation without first switching off the power at the consumer unit and removing the relevant circuit fuse or locking off the miniature circuit breaker (MCB).
- Always unplug a portable appliance or light before doing any work on it.
- Always double-check all your work (especially connections) before you turn the electricity on again.
- Always use the correct tools for an electrical job, and use good-quality equipment and materials.
- Fuses are vital safety devices. Never fit one that's rated too highly for the circuit it is to protect – and never be tempted to use any other type of wire or metal strip in place of proper fuses or fuse wire.
- Wear rubber-soled shoes when you're working on an electrical installation.

Colour coding

For purpose of identification, the coverings of live, neutral and earth wires in flex and mains cables are colour-coded. The live wires are brown and the neutral wires blue.

When an earth wire is included in a piece of flex, it is coded with a green-and-yellow covering. In a mains cable, the earth is a bare copper wire sandwiched between the insulated live and neutral wires. Whenever the bare earth wire is exposed for linking to socket outlets or light fittings, it should be covered with a green-and-yellow sleeve.

Until recently the live and neutral wires in mains cables were coded with other colours. Live wires had a red covering, and neutral wires a black covering. There is no reason to replace old colour-coded cables – but whenever you need to join new cable to old, remember to connect the brown wire to the old red one, and the blue wire to the old black one.

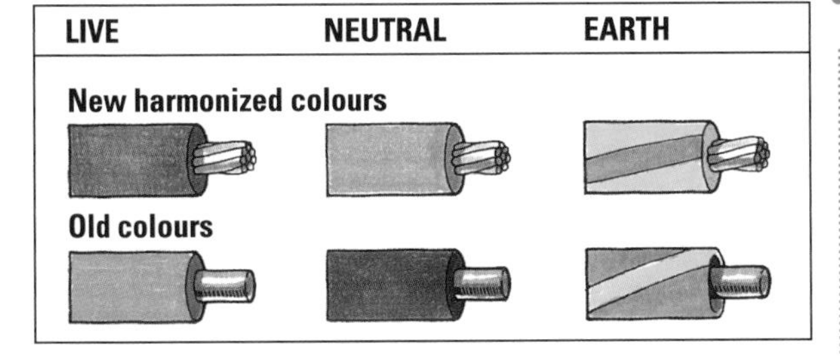

Identifying conductors
The insulation used to cover the conductors in electrical cable and flex is colour-coded to indicate live, neutral and earth.

Is the power off?

Having turned off the power, you can make doubly sure that an accessory is safe to work on by using an electronic mains-voltage tester to check whether terminals or wires are live, before you tamper with them. Always make sure the tester itself is functioning properly before and after you use it, by testing it on a circuit you know to be live.

Following the maker's instructions, put one probe on the neutral terminal and the other on the live terminal; if the indicator lights up, the circuit is live. If it doesn't illuminate, test again – this time between the earth terminal and each of the live and neutral terminals. If the indicator still doesn't light up, you can assume the circuit is not live – provided you have checked the tester.

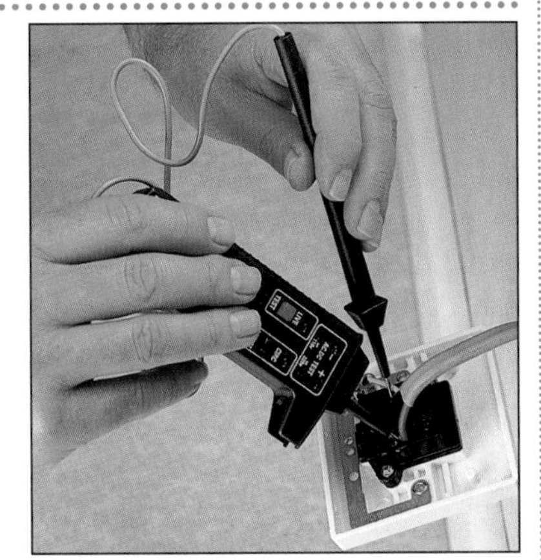

Using an electronic tester
Touch the neutral terminal with one probe and the live terminal with the other. The circuit is live if the indicator illuminates.

FUSES AND CIRCUIT BREAKERS

A conductor will heat up if an unusually powerful current flows through it. This can damage electrical equipment and create a serious fire risk if it is allowed to continue in any part of the domestic wiring system. As a safeguard, weak links are included in the wiring to break the circuit before the current reaches a dangerously high level.

A very common form of protection is a fuse, a thin wire that's designed to break the circuit by melting at a specific current. This varies according to the part of the system that the fuse is protecting – an individual appliance, a single power or lighting circuit, or the entire domestic wiring system.

Alternatively, a special switch called a circuit breaker is used that trips and cuts off the current as soon as an overload on the wiring is detected.

A fuse will 'blow' in the following circumstances:

- If too many appliances are operated on a circuit simultaneously, then the excessive demand for electricity will blow the fuse in that circuit.
- If the current reroutes to earth due to a faulty appliance, the flow of power increases in the circuit and blows the fuse (this is known as an earth fault).

WARNING: The original fault must be dealt with before the fuse is replaced.

Measuring electricity

Watts measure the amount of power used by an appliance when working. The wattage of an electrical appliance is normally marked on its casing.

One thousand watts (1000W) equal one kilowatt (1kW).

Amps measure the flow of current that is necessary to produce the required wattage for an appliance.

Volts measure the 'pressure' provided by the electricity company. This drives the current along the conductors to the various outlets. In this country 230 volts (formerly 240) is standard.

If you know two of these measurements, you can determine the other one:

$\frac{\text{Watts}}{\text{Volts}}$ = Amps	Amps x Volts = Watts
Use this method to determine what kind of fuse or flex is safe.	Indicates how much power is needed to operate an appliance.

Bathroom safety

Because water is such a highly efficient conductor of electric current, water and electricity form a very dangerous combination. For this reason, in terms of electricity bathrooms are potentially the most dangerous areas in your home. Where there are so many exposed metal pipes and fittings, combined with wet conditions, regulations must be stringently observed if fatal accidents are to be avoided.

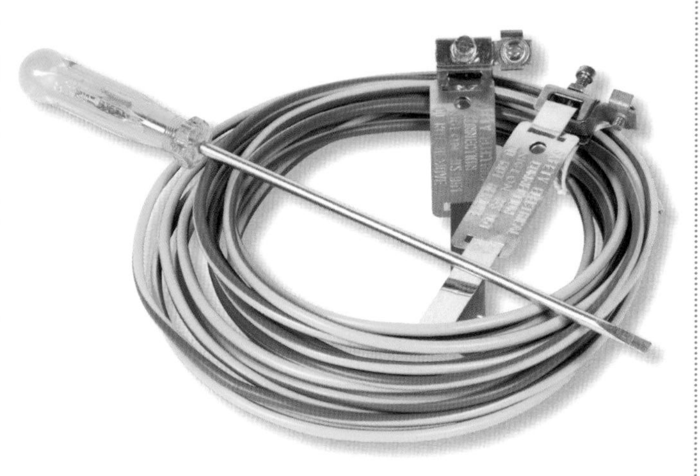

GENERAL SAFETY

- Sockets must not be fitted in a bathroom – except for special shaver sockets that conform to BS EN 60742 Chapter 2, Section 1.
- The IEE Wiring Regulations stipulate that light switches in bathrooms must be outside zones 0 to 3 (see opposite). The best way to comply with this requirement is to fit only ceiling-mounted pull-cord switches.
- Any bathroom heater must comply with the IEE Wiring Regulations.
- If you have a shower in a bedroom, it must be not less than 3m (9ft 11in) from any socket outlet, which must be protected by a 30 milliamp RCD.
- Light fittings must be well out of reach and shielded – so fit a close-mounted ceiling light, properly enclosed, rather than a pendant fitting.
- Never use a portable fire or other electrical appliance, such as a hairdryer, in a bathroom – even if it is plugged into a socket outside the room.

• RCD protection
When installing any electrical appliance in a bathroom, the circuit should be protected by a 30 milliamp RCD.

Supplementary bonding

In any bathroom there are many non-electrical metallic components, such as metal baths and basins, supply pipes to bath and basin taps, metal waste pipes, radiators, central-heating pipework and so on – all of which could cause an accident during the time it would take for an electrical fault to blow a fuse or operate a miniature circuit breaker (MCB). To ensure that no dangerous voltages are created between metal parts, the Wiring Regulations stipulate that all these metal components must be connected one to another by a conductor which is itself connected to a terminal on the earthing block in the consumer unit. This is known as supplementary bonding and is required for all bathrooms – even when there is no electrical equipment installed in the room, and even though the water and gas pipes are bonded to the consumer's earth terminal near the consumer unit.

When electrical equipment such as a heater or shower is fitted in a bathroom, that too must be supplementary-bonded by connecting its metalwork – such as the casing – to the nonelectrical metal pipework, even though the appliance is connected to the earthing conductor in the supply cable.

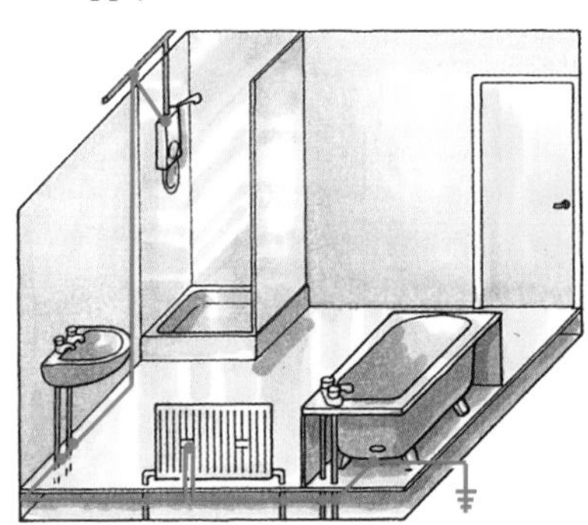

Supplementary bonding in a bathroom

Making the connections

The Wiring Regulations specify the minimum size of earthing conductor that can be used for supplementary bonding in different situations, so that large-scale electrical installations can be costed economically. In a domestic environment, use $6mm^2$ single-core cable insulated with green-and-yellow PVC for supplementary bonding. This is large enough to be safe in any domestic situation. For a neat appearance, plan the route of the bonding cable to run from point to point behind the bath panel, under floorboards, and through basin pedestals. If necessary, run the cable through a hollow wall or under plaster, like any other electrical cable.

Connecting to pipework

An earth clamp **(1)** is used for making connections to pipework. Clean the pipe locally with wire wool to make a good connection between the pipe and clamp, and scrape or strip an area of paintwork if the pipe has been painted.

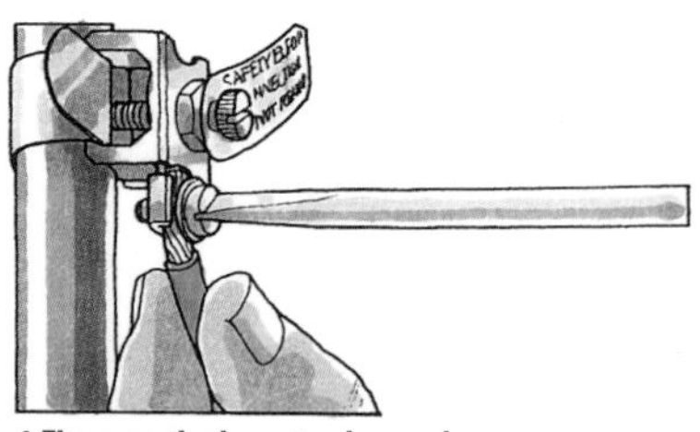

1 Fit an earth clamp to pipework

Connecting to a bath or basin

Metal baths or basins are made with an earth tag. Connect the earth cable by trapping the bared end of the conductor under a nut and bolt with metal washers **(2)**. Make sure the tag has not been painted or enamelled.

If an old metal bath or basin has not been provided with an earth tag, drill a hole through the foot of the bath or through the rim at the back of the basin; and connect the cable with a similar nut and bolt, with metal washers.

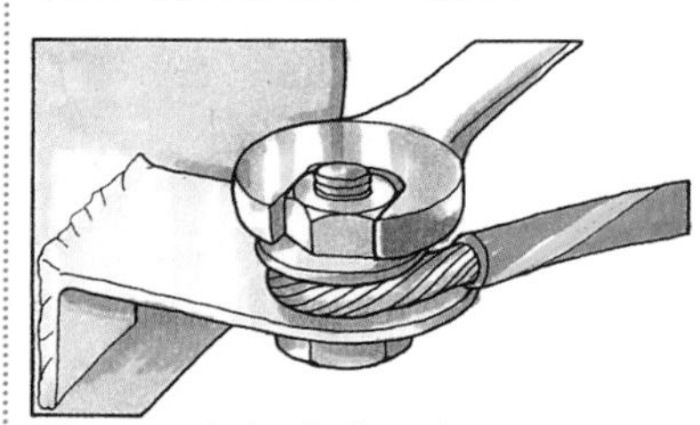

2 Connect to bath or basin earth tag

Connecting to an appliance

Simply connect the earth cable to the terminal provided in the electrical appliance **(3)** and run it to a clamp on a metal supply pipe nearby.

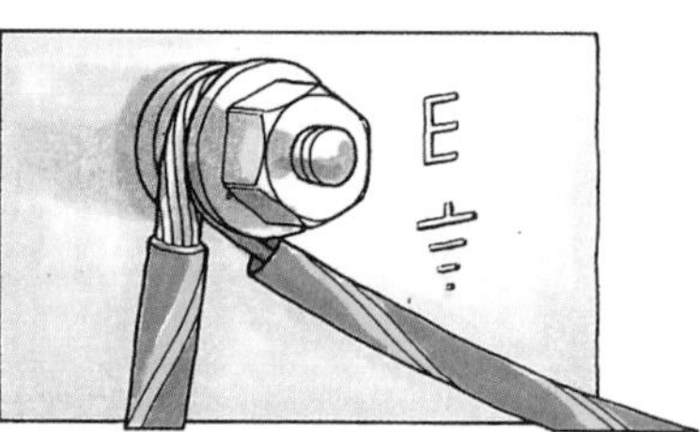

3 Fix to the earth terminal in an appliance
The appliance's own earth connection may share the same terminal.

WARNING

Have your supplementary bonding tested by a qualified electrician. Unless you consider yourself fully competent to do the work, have supplementary bonding installed by a professional.

SEE ALSO: Building Regulations 8, Bonding to earth 16–17, PME 17, Cables 22, Running cable 23–5, Bathroom heaters 35, Shaver sockets 36, Electric shower 47, Close-mounted lights 49, Ceiling switch 51, 53

Zones for bathrooms

Within a room containing a bath or shower, the IEE Wiring Regulations define areas, or zones, where specific safety precautions apply. The regulations also describe what type of electrical appliances can be installed in each zone, and the routes cables must take in order to serve those appliances. There are special considerations for extra-low-voltage equipment with separated earth; this is best left to a qualified electrician.

The four zones

Any room containing a bathtub or shower is divided into four zones. Zone 0 is the interior of the bathtub or shower tray – not including the space beneath the tub, which is covered by other regulations (see top right). Zones 1 to 3 are specific areas above and all round the bath or shower, where only specified electrical appliances and their cables may be installed. Wiring outside these areas must conform to the IEE Wiring Regulations, but no specific 'zone' regulations apply.

ZONE	LOCATION	PERMITTED
Zone 0	Interior of the bathtub or shower tray.	No electrical installation.
Zone 1	Directly above the bathtub or shower tray, up to a height of 2.25m (7ft 5in) from the floor. (See also top right.)	Instantaneous water heater. Instantaneous shower. All-in-one power shower, with a suitably waterproofed integral pump. The wiring that serves appliances within the zone.
Zone 2	Area within 0.6m (2ft) horizontally from the bathtub or shower tray in any direction, up to a height of 2.25m (7ft 5in) from the floor. The area above zone 1, up to a height of 3m (9ft 11in) from the floor.	Appliances permitted in zone 1. Light fittings. Extractor fan. Space heater. Whirlpool unit for the bathtub. Shaver socket to BS EN 60742 Chapter 2, Section 1. The wiring that serves appliances within the zone and any appliances in zone 1.
Zone 3	Up to 2.4m (7ft 11in) outside zone 2, up to a height of 2.25m (7ft 5in) from the floor. The area above zone 2 next to the bathtub or shower, up to a height of 3m (9ft 11in) from the floor.	Appliances permitted in zones 1 and 2. Any fixed electrical appliance (a heated towel rail, for example) that is protected by a 30 milliamp RCD. The wiring that serves appliances within the zone and any appliances in zones 1 and 2.

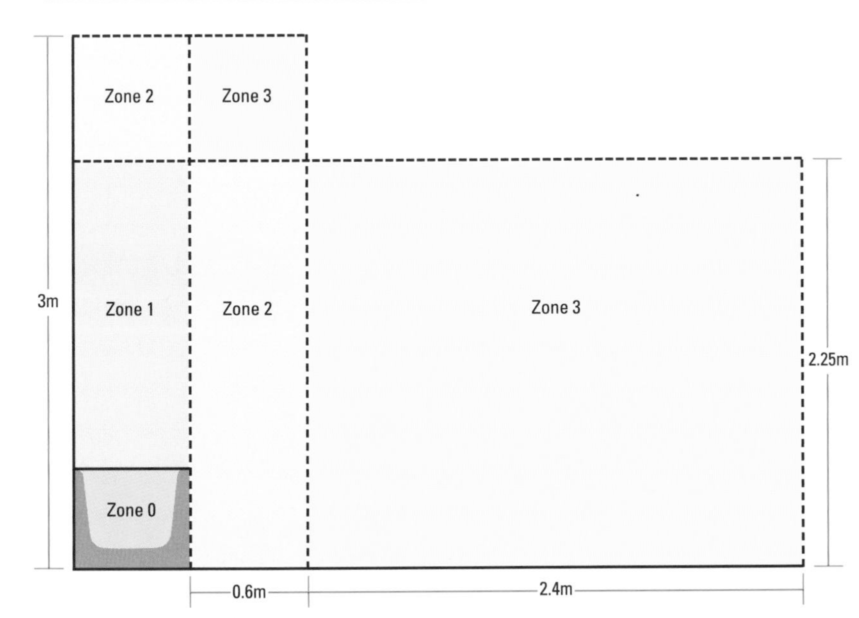

Zones within a room containing a bath or shower

UNDER THE BATH

The space under a bathtub is designated as zone 1 if it is accessible without having to use a tool – that is, if there is no bath panel or if the panel is attached with magnetic catches or similar devices that allow the panel to be detached without using a tool of some kind. If, however, the panel is screw-fixed – so that it can only be removed with the aid of a screwdriver – then the enclosed space beneath the bath is considered to be outside all zones.

Supplementary bonding

In bathrooms, nonelectrical metallic components must be bonded to earth (see opposite). In zones 1, 2 and 3, this supplementary bonding is required to all pipes, any electrical appliances and any exposed metallic structural components of the building. This does not include window frames, unless they are themselves connected to metallic structural components.

Supplementary bonding is not required outside the zones. And in the special case of a bedroom containing a shower cubicle, supplementary bonding can also be omitted from zone 3.

● Cable runs
You are not permitted to run electrical cables that are feeding a zone through another zone designated with a lower number. This includes cables buried in the plaster or concealed behind other wallcoverings.

Switches

Electrical switches, including ceiling-mounted switches operated by a pull cord, must be situated outside the zones. The only exceptions are those switches and controls incorporated in appliances suitable for use in the zones.

If the bathroom ceiling is higher than 3m (9ft 11in), ceiling-mounted pull-cord switches can be mounted anywhere. However, if the ceiling height is between 2.25 and 3m (7ft 5in and 9ft 11in), pull-cord switches must be mounted at least 0.6m (2ft) – measured horizontally – from the bathtub or shower cubicle. If the ceiling is lower than 2.25m (7ft 5in), switches can only be mounted outside the room.

● 13amp sockets
In the special case of a bedroom containing a shower cubicle, socket outlets are permitted in the room, but only outside the zones, and the circuit that feeds the sockets must be protected by a 30 milliamp RCD.

IP coding

Electrical appliances installed in zones 1 and 2 must be manufactured with suitable protection against splashed water. This is designated by the code IPX4 (the letter X is sometimes replaced with a single digit). Any number larger than four is also acceptable as this indicates a higher degree of waterproofing. If in doubt, check with your supplier that the appliance is suitable for its intended location.

IP coding
Suitable equipment may be marked with the symbol shown above.

SEE ALSO: Wiring heaters 35, Wiring a shower unit 47, New switches 53

Simple replacements

Twisted twin flex
This is similar to parallel twin flex (above right), but the insulated conductors are twisted together for extra strength. It was once used to support hanging light fittings, but nowadays must be replaced with a two-core sheathed flex when wiring pendant lights. Also, any old rubber-insulated flex with braided-cotton covering, which is still found in some homes, should be replaced.

You can carry out many repairs and replacements without having to concern yourself with the wiring system installed in your home. Many light fittings and appliances are supplied with electricity by means of flexible cords that plug into the system – so provided that they have been disconnected, there can be no risk of getting an electric shock while working on them.

Flexible cord (flex)

All portable appliances and some of the smaller fixed ones, as well as pendant and portable light fittings, are connected to your home's permanent wiring system by means of conductors in the form of flexible cord, normally called 'flex'.

Each of the conductors in any type of flex is made up of numerous fine wires twisted together, and each conductor is insulated from the others by a covering of plastic insulation. So that the conductors can be identified easily, the insulation is usually colour-coded (brown = live; blue = neutral; and green-and-yellow = earth).

Further protection is provided on most flexible cords in the form of an outer sheathing of insulating material enclosing the inner conductors.

Heat-resistant flex is available for enclosed light fittings and appliances with surfaces that become hot.

WARNING

Never attempt to carry out electrical repairs without first unplugging the appliance or switching off the power supply at the consumer unit.

Types of electrical flex

Parallel twin

Parallel twin flex has two conductors, insulated with PVC (polyvinyl chloride), running side by side. The insulation material is joined between the two conductors along the length of the flex. This kind of flex should only be used for wiring audio-equipment speakers. One of the conductors will be colour-coded for identification.

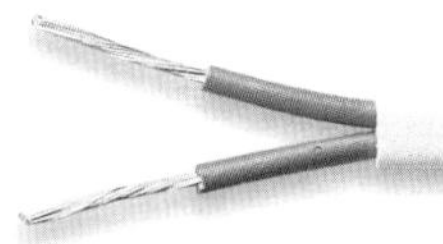

Flat twin sheathed

Flat twin sheathed flex has colour-coded live and neutral conductors inside a PVC sheathing. This flex is used for double-insulated light fittings and small appliances.

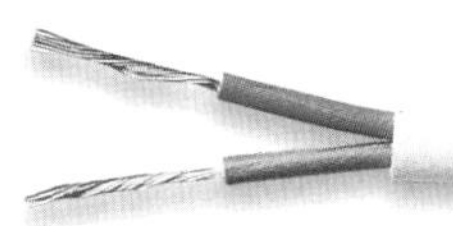

Two-core circular sheathed

This has colour-coded live and neutral conductors inside a PVC sheathing that is circular in its cross section. It is used for wiring certain pendant lights and some double-insulated appliances.

Three-core circular sheathed

This is like two-core circular sheathed flex, but it also contains an insulated and colour-coded earth wire. This flex is perhaps the most commonly used for all kinds of appliances. A special high-temperature flex is available for connecting immersion heaters, storage heaters and similar appliances.

Unkinkable braided

This flex is used for appliances such as kettles and irons, which are of a high wattage and whose flex must stand up to movement and wear. The three rubber-insulated conductors, plus the textile cords that run parallel with them, are all contained in a rubber sheathing that is bound outside with braided material. This type of flex can be wound round the handle of a cool electric iron.

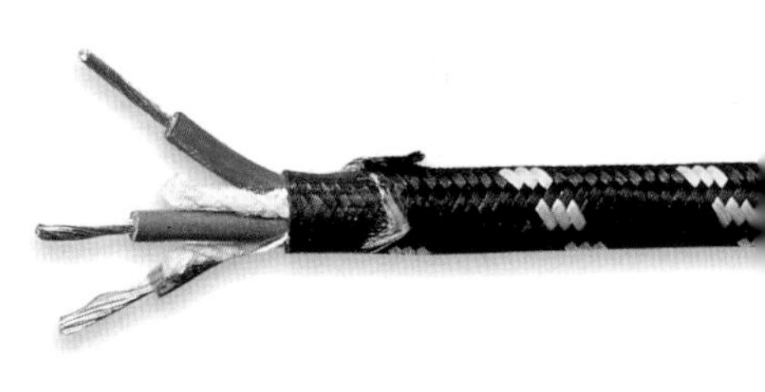

Coiled flex

A coiled flex that stretches and retracts can be a convenient way of connecting a portable lamp or appliance.

SEE ALSO: Colour coding 9, Switching off 16

Connecting flexible cord

Although the spacing of terminals in plugs and appliances varies, the method of stripping and connecting the flex is the same.

Stripping the flex

Crop the flex to length **(1)**. Slit the sheath lengthwise with a sharp knife **(2)**, being careful not to cut into the insulation covering the individual conductors. Peel the sheathing away from the conductors, then fold it back over the knife blade and cut it off **(3)**.

Separate the conductors, crop them to length and, using wire strippers, remove about 12mm (½in) of insulation from the end of each one **(4)**.

Divide the conductors of parallel twin flex by pulling them apart before exposing their ends with wire strippers.

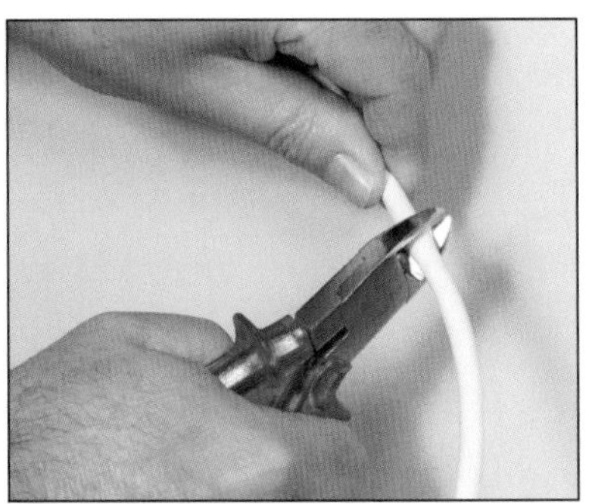

1 Crop the flex to length

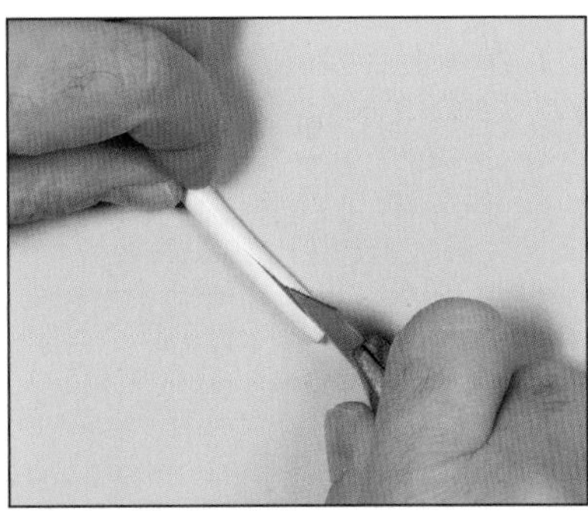

2 Slit sheathing lengthwise

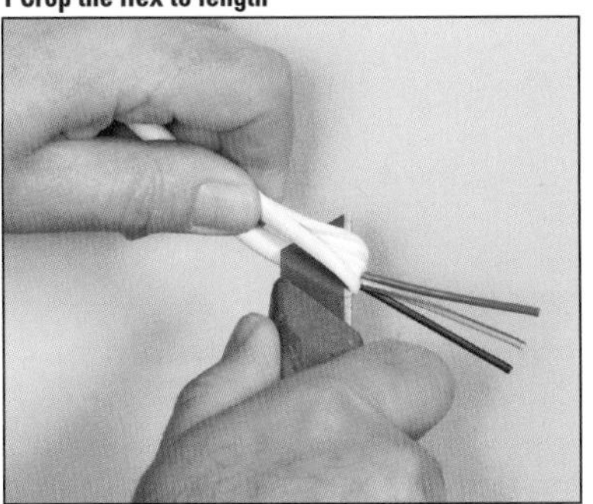

3 Fold sheathing over the blade and cut it off

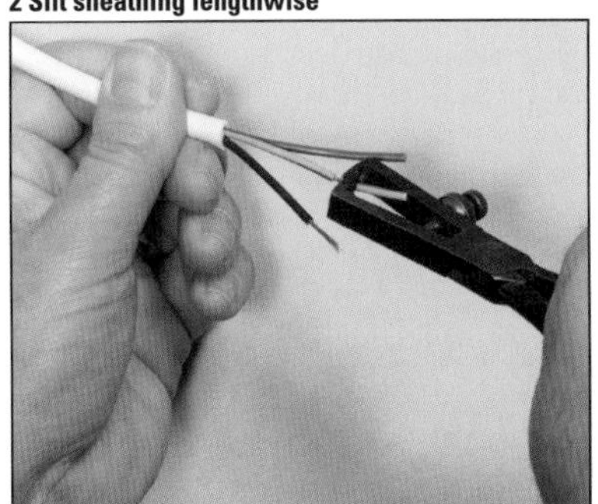

4 Strip insulation from conductors

Connecting conductors

Twist together the individual filaments of each conductor to make them neat.

If the plug or appliance has post-type terminals, fold the bared end of wire **(1)** before pushing it into the hole. Make sure the insulation butts against the post and that all the wire filaments are enclosed within the terminal. Then tighten the clamping screw, and pull gently on the wire to make sure it is held quite firmly.

When you're connecting to clamp-type terminals, wrap the bared wire round the threaded post clockwise **(2)**, then screw the clamping nut down tight onto the conductor. After tightening the nut, check that the conductor is held securely.

1 Post terminal

2 Clamp terminal

CHOOSING A FLEX

Not only is the right type of flex for the job important; the size of its conductors must suit the amount of current that will be used by the appliance.

Flex is rated according to the area of the cross section of its conductors, $0.5mm^2$ being the smallest for normal domestic wiring. The flex size required is determined by the flow of current that it can handle safely. Excessive current will make a conductor overheat – so the size of the flex must be matched to the power (wattage) of the appliance that it is feeding.

Manufacturers often fit $1.25mm^2$ flex to appliances of less than 3000W (3kW), since it is safer to use a larger conductor than necessary if a smaller flex might be easily damaged. It is advisable to adopt the same procedure when replacing flex.

• **Flexible cord for immersion heaters**
Because they generate relatively high background temperatures, 3kW immersion heaters are wired with $2.5mm^2$ heat-resistant flex (see WIRING AN IMMERSION HEATER).

Conductor	Current rating	Appliance
$0.5mm^2$	3amp	Light fittings up to 720W
$0.75mm^2$	6amp	Light fittings and appliances up to 1440W
$1.0mm^2$	10amp	Appliances up to 2400W
$1.25mm^2$	13amp	Appliances up to 3120W
$1.5mm^2$	15amp	Appliances up to 3600W
$2.5mm^2$	20amp	Appliances up to 4800W

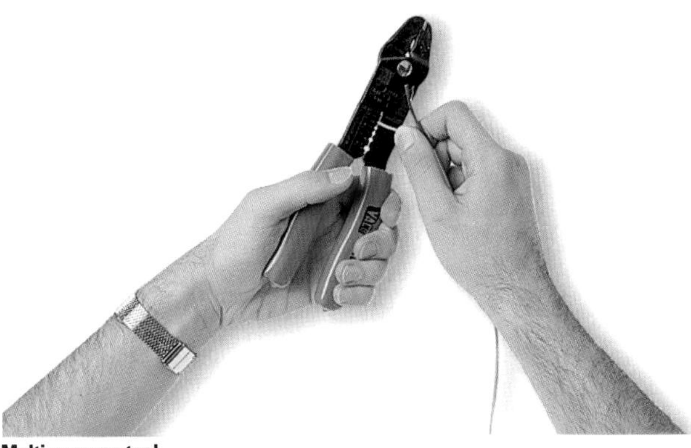

Multi-purpose tool
This tool will crop and strip any size of cable or flex.

SEE ALSO: Measuring electricity 9, Immersion heaters 38–9, Wire strippers 67

Extending flexible cord

When you plan the positions of socket outlets, try to ensure there will be enough, all conveniently situated, so that it's never necessary to extend the flexible cord of a table lamp or other appliance. But if you do find that a flex will not reach a socket, extend it so that it is not stretched taut, which may cause an accident.

Never be tempted to join two lengths of flex by twisting the bared ends of wires together, even if you bind them with insulating tape. People often do this as a temporary measure then neglect to make a proper connection later – which can have fatal consequences.

Flex connectors

If possible, fit a longer flex, wiring it into the appliance itself. But if you can't do this or don't want to dismantle the appliance, use a flex connector. There are two-terminal and three-terminal connectors, which you must match to the type of flex you are using. Never join two-core flex to three-core flex.

Strip off just enough sheathing for the conductors to reach the terminals, and make sure the sheathed part of each cord can be secured under the cord clamp at each end of the connector.

Crop the conductors to length, then strip and connect the conductors – connecting the live conductor to one of the outer terminals, the neutral to the other, and the earth wire (if present) to the central terminal. Make sure that matching conductors from both cords are connected to the same terminals, then tighten the cord clamps and screw the cover in place.

In-line switches
If you plan to fit a longer continuous length of flex you can install an in-line switch that will allow you to control the appliance or light fitting from some distance away – a great advantage for the elderly and people confined to bed. Some in-line switches are luminous.

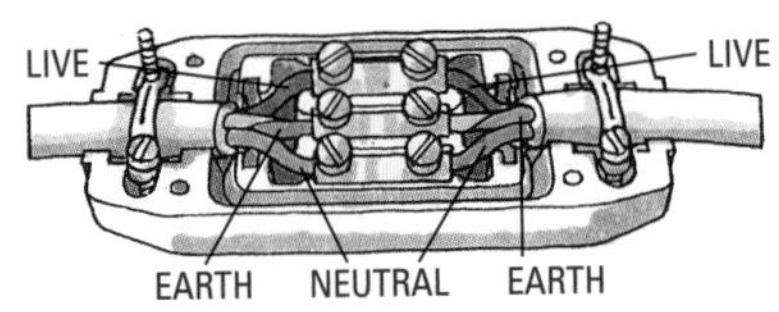

Wiring a flex connector

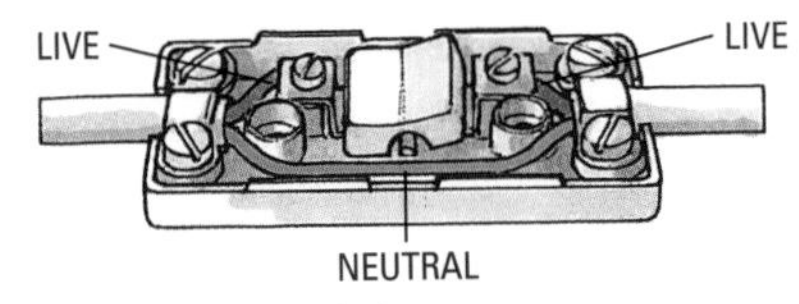

Wiring an in-line switch

Extension leads

If you fit a long flex to a power tool, it will inevitably become tangled and one of the conductors will eventually break, perhaps causing a short circuit. The solution is to buy an extension lead or make one yourself.

The best type of extension lead to be had commercially is wound on a drum. There are 5amp ones – but it's safer to buy one with a 13amp rating, so you can run a wider range of equipment without danger of overloading. If you use such a lead while it is wound on the drum it may overheat, so develop the habit of unwinding it fully each time you use it. The drums of these leads have a built-in 13amp socket to take the plug of the appliance; the plug at the end of the lead is then connected to a wall socket.

You can make an extension lead from a length of 1.5mm^2 three-core flex with a standard 13amp plug on one end and a trailing socket on the other. Use those with unbreakable rubber casings. A trailing socket is wired in a similar way to a 13amp plug (see opposite). Its terminals are marked to indicate which conductors to connect to them.

'Multi-way' trailing sockets will take several plugs and are ideal for hi-fi systems or computers with individual components that need to be connected to the mains supply. Using a multi-way socket, the whole system is supplied from a single plug in the wall socket.

You can also extend a lead by using a lightweight two-part flex connector. One half has three pins that fit into the other half of the connector.

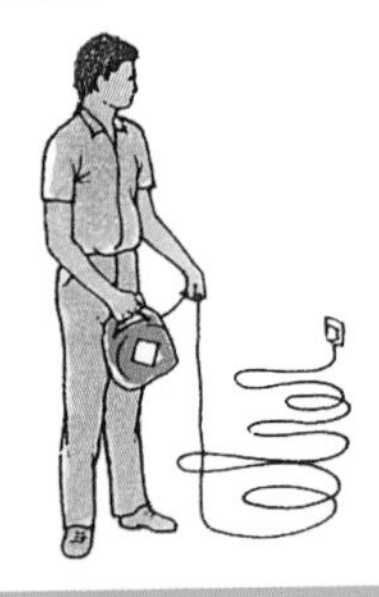
Unwind the lead
Always fully unwind a 13amp extension lead before you plug in an appliance rated at 1kW or more.

TYPES OF FLEX EXTENDER

Below are illustrated four of the devices available for extending the flexible cords of electrical appliances.

Drum-type extension lead

13amp plug and trailing socket

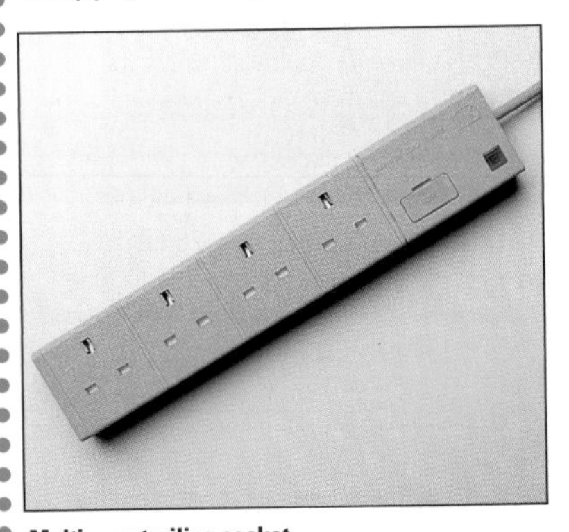
Multi-way trailing socket

Two-part flex connector
When wiring a two-part flex connector, never attach the part with the pins to the extension lead. The exposed pins will become live – and dangerous – when the lead is plugged into the socket.

SEE ALSO: Flex 12, Connecting flex 13, Positioning sockets 28

Three-pin plugs

All sorts of plug were once in use in this country, but today standard 13amp square-pin plugs are used for all portable appliances and light fittings. They are available with rigid plastic or unbreakable rubber casings. Some plugs have neon indicators to show when they are live, and some have pins insulated for part of their length to prevent the user getting a shock from a plug pulled partly from the socket. Use only plugs marked BS 1363.

Fuses for plugs

Square-pin plugs have a small cartridge fuse to protect the appliance. Use a 3amp (red) fuse for appliances of up to 720W, and a 13amp (brown) fuse for those of 720 to 3000W (3kW). There are also 2, 5 and 10amp fuses, but these are less often used in the home.

Wiring a 13amp plug

Loosen the large screw between the pins and remove the cover. Position the flex on the open plug to gauge how much sheathing to remove (remember that the cord clamp must grip sheathed flex, not the conductors).

Strip the sheathing and position the flex on the plug again, so that you can cut the conductors to the right length. These should take the most direct routes to their terminals and lie neatly within the channels of the plug.

Strip and prepare the ends of the wires, then secure each to its terminal. If you are using two-core flex, wire to the live and neutral terminals, leaving the earth terminal empty.

Tighten the cord clamp to grip the end of the sheathing and secure the flex (one type of plug has a sprung cord grip that tightens if the flex is pulled hard). Check that a fuse of the correct rating is fitted, then replace the plug's cover and tighten up the screw.

Round-pin plugs

Old round-pin sockets will only take round-pin plugs, which are not fused. Use 2amp plugs for lighting only; 5amp plugs for appliances of up to 1kW; and 15amp plugs for appliances between 1kW and 3kW. Have your wiring upgraded as soon as possible, so you can use modern fused square-pin plugs.

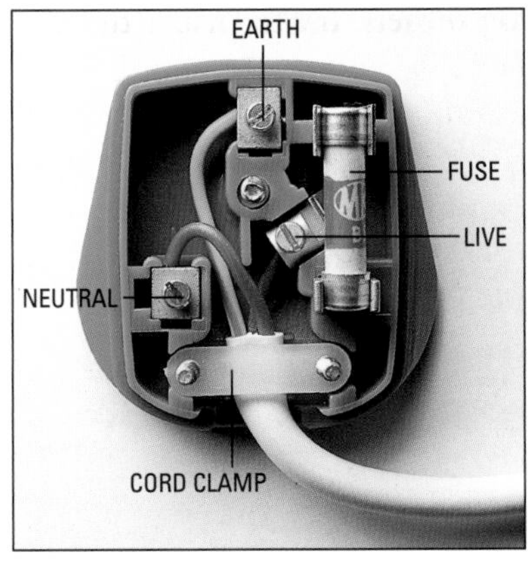

Post-terminal plug

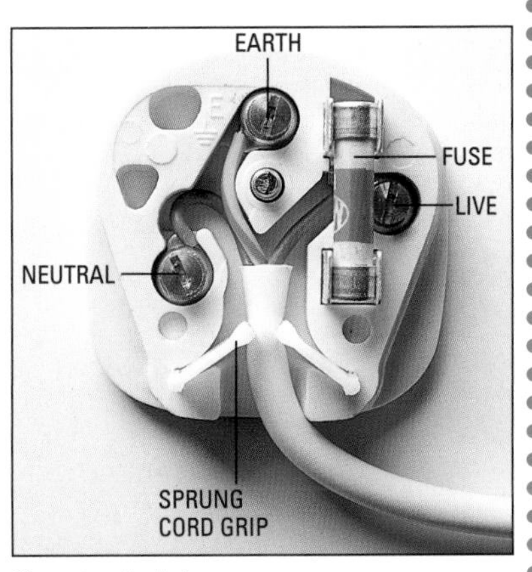

Clamp-terminal plug

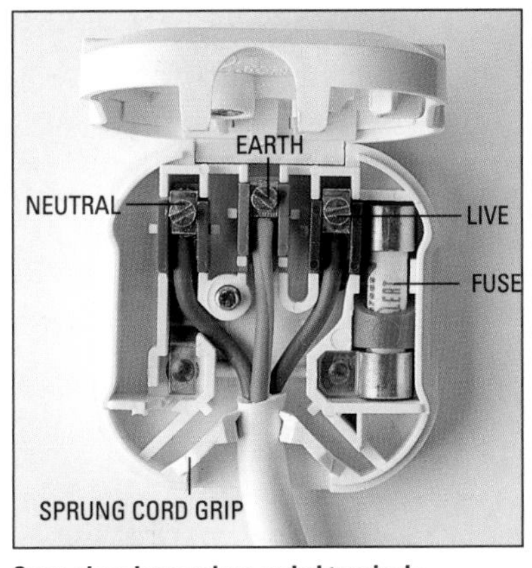

Some plugs have colour-coded terminals

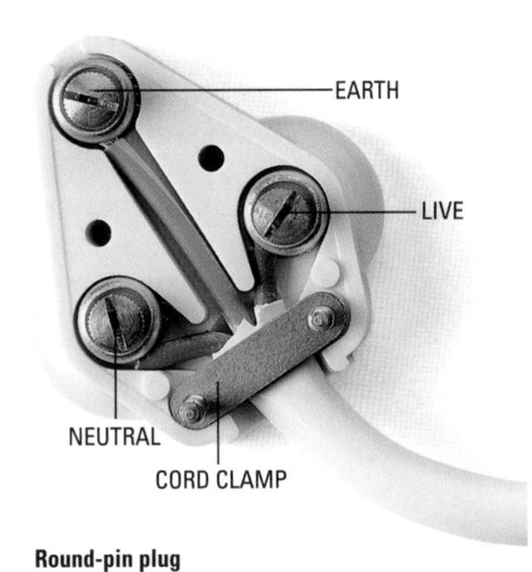

Round-pin plug

REPLACING A PENDANT LAMPHOLDER

Because they are usually well out of reach, damaged pendant lampholders often go unnoticed. So check their condition from time to time, and replace any that look suspect before they become dangerous.

Because pendant lampholders hang on flex from the ceiling, they are in a stream of hot air rising from the bulb. In time this tends to make plastic holders brittle and more easily cracked or broken.

On a metal lampholder, the earth wire can become detached or corroded so that the fitting is no longer safe.

Types of lampholder

Plastic lampholders are the most common type. These have a threaded skirt that screws onto the actual holder (the part that takes the bulb). Some have an extended skirt. If you are going to use a close-fitting or badly ventilated shade, fit a heat-resistant version. Plastic holders are designed to take two-core flex only. Never fit one on a three-core flex, as there is no place to attach the earth wire.

Metal lampholders are similar in construction, but they must be wired with three-core flex so that they can be connected to earth. Never fit a metal lampholder in a bathroom – and never attach one to a two-core flex, which lacks an earth conductor.

Fitting a lampholder

Before commencing work, remove the circuit fuse or remove (or lock off) the circuit breaker from the consumer unit so that no-one can turn the power on. Unscrew the old holder's cap – or the retaining ring if it's a metal one – and slide it up the flex to expose the terminals. Loosen their screws and pull the wires out. If some wires are broken or brittle, cut back slightly to expose sound wires before fitting the new holder.

Slide the cap of the new fitting up the flex and attach it temporarily with adhesive tape. Fit the live wire into one of the terminals, and the neutral wire into the other one. Then loop the conductors round the supporting lugs of the holder, to take the weight off the terminals, and screw the cap down.

On a metal holder, pass the earth wire through the hole in the cap before you secure it. Connect the earth wire to the earth terminal, then secure the cap with the retaining ring.

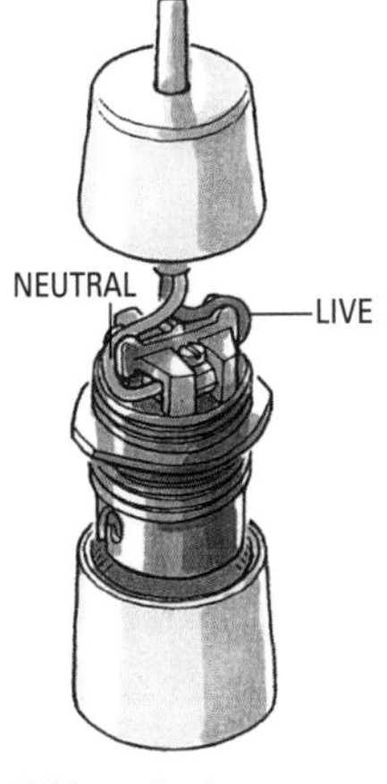

Wiring a plastic pendant lampholder

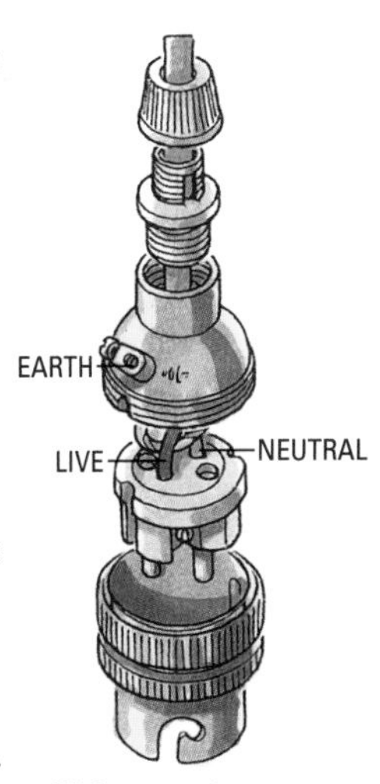

Wiring a metal pendant lampholder

SEE ALSO: Flex 12, Connecting flex 13, Switching off 16

Main switch equipment

Electricity flows because of a difference in 'pressure' between the live wire and the neutral one, and this difference in pressure is measured in volts. Domestic electricity in this country is supplied as alternating current, at 230 volts, by way of the electricity company's main service cable. This normally enters your house underground, although in some areas electricity is distributed by overhead cables.

The service head

The main cable terminates at the service head, or 'cutout', which contains the service fuse. This fuse prevents the neighbourhood's supply being affected if there should be a serious fault in the circuitry of your house. Cables connect the cutout to the meter, which registers how much electricity you consume. Both the meter and cutout belong to the electricity company and must not be tampered with. The meter is sealed in order to disclose interference.

If you use cheap night-time power for storage heaters and hot water, a time switch will be supplied by the electricity company.

• Cross-bonding cable sizes
Single-core cables are used to cross-bond gas and water pipes to earth. An electrician can calculate the minimum size for these cables, but for any single house or flat it is safe to use $10mm^2$ cable. (See also PME opposite).

Consumer unit

Electricity is fed to and from the consumer unit by 'meter leads', thick single-core insulated-and-sheathed cables made up of several wires twisted together. The consumer unit is a box that contains the fuseways that protect the individual circuits in the house. It also incorporates the main isolating switch, which you operate when you need to cut off the supply of power to the whole house.

In a house where several new circuits have been installed over the years, the number of circuits may exceed the number of fuseways in the consumer unit. If so, an individual switchfuse unit – or more than one – may have been mounted alongside the main unit. Switchfuse units comprise a single fuseway and an isolating switch; they, too, are connected to the meter by means of meter leads.

If your home is heated by off-peak storage heaters, then you will have an Economy 7 meter and a separate consumer unit for the heater circuits.

• The main isolating switch
Not all main isolating switches operate the same way. Before you need to use it, check to see whether the main switch on your consumer unit has to be in the up or down position for 'off'.

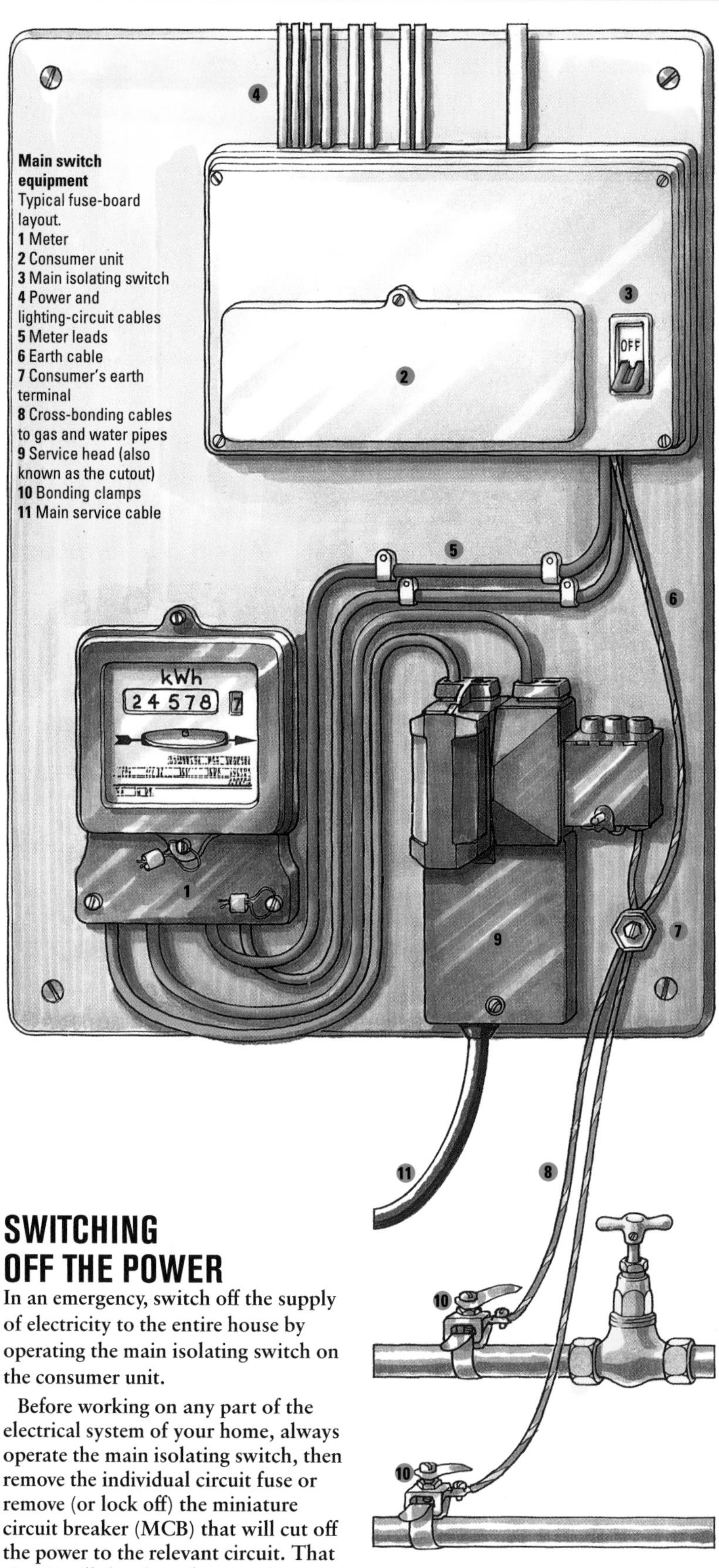

Main switch equipment
Typical fuse-board layout.
1 Meter
2 Consumer unit
3 Main isolating switch
4 Power and lighting-circuit cables
5 Meter leads
6 Earth cable
7 Consumer's earth terminal
8 Cross-bonding cables to gas and water pipes
9 Service head (also known as the cutout)
10 Bonding clamps
11 Main service cable

SWITCHING OFF THE POWER

In an emergency, switch off the supply of electricity to the entire house by operating the main isolating switch on the consumer unit.

Before working on any part of the electrical system of your home, always operate the main isolating switch, then remove the individual circuit fuse or remove (or lock off) the miniature circuit breaker (MCB) that will cut off the power to the relevant circuit. That circuit will then be safe to work on, even if you restore the power to the rest of the house by operating the main switch again.

SEE ALSO: Cheaper electricity 6, Consumer units 18, Circuit breakers 19, Fuses 19, Switchfuse unit 38, Storage heaters 40–1

Earthing systems

The earthing system

All of the individual earth conductors of the various circuits in the house are connected to a metal earthing block in the consumer unit. A single cable with a green-and-yellow covering runs from this earthing block to the consumer's earth terminal, which is mounted next to the cutout. In most urban houses a connection is provided from inside the cutout to an external earth-connection block, which is also wired to the consumer's earth terminal. This provides an effective path to earth, as it allows the current to pass along the sheath of the main service cable to the electricity company's substation, where it is solidly connected to earth.

In the past most domestic electrical systems were earthed to the cold-water supply, so earth-leakage current passed out along the metal water pipes into the ground in which they were buried. But nowadays more and more water systems use nonconductive, nonmetallic pipes and fittings. As a result, such a means of earthing is no longer reliable.

Despite this, you will find that your gas and water pipework is connected to the consumer's earth terminal. This ensures that the water and gas piping systems are cross-bonded, so earth-leakage current passing through either system will run without hindrance to the main earth without producing dangerously high voltages. The cross-bonding clamps must be as close as possible to the point where the pipes enter the house, but on the consumer's side (within 600mm) of the stopcock or gas meter.

Bonding clamp
This type of clamp (BS 951) is used to make connections to gas and water pipes. It must not be removed under any circumstances.

PME

The electricity company sometimes provides a different method of earthing the system, called 'protective multiple earth' (PME), by which earth-leakage current is fed back to the substation along the neutral return wire, and so to earth. Regulations regarding the earthing of this system are particularly stringent. With PME, cross-bonding cables to gas and water services are sometimes required to be larger. Check this with the electricity company.

RCDs

Although the local electricity company normally provides effective earthing for the electrical system of your home, safe earthing is the consumer's own responsibility. With this in mind, it is worth installing a residual current device (RCD) into the house circuitry.

When conditions are normal, the current flowing out through the neutral conductor is exactly the same as that flowing in through the live one. Should there be an imbalance between the two caused by an earth leakage, the RCD will detect it immediately and isolate the circuitry.

An RCD can be either installed as a separate unit or incorporated into the consumer unit together with the main isolating switch.

A residual current device is sometimes referred to as a residual current circuit breaker (RCCB). It was formerly known as an ELCB, or earth-leakage circuit breaker.

A separate unit containing an RCD

RECOGNIZING AN OLD FUSE BOARD

Domestic wiring systems were once very different from the ones used today. Besides lighting, water-heating and cooker circuits, each socket outlet had its own circuit and fuse, while further circuits would be installed from time to time as the needs of the household changed. Consequently, an old house may have a mixture of 'fuse boxes' attached to the fuse board, along with the meter.

You may find that the wiring itself is haphazard and badly labelled, with the serious danger that you may not safely isolate a circuit you're going to work on. Furthermore, you will not be able to tell whether a particular fuse is correctly and safely rated unless you know what type of circuit it is protecting.

Arrange for an inspection

If your home still has such an old-style fuse board, have it inspected and tested by a qualified electrician before you attempt to work on any part of the system. He or she can advise you as to whether your installation needs to be replaced with a modern consumer unit – and if it does prove to be in good working condition, he or she can label the various circuits clearly to help you in the future.

An old-fashioned fuse board
This type of installation is out of date. A professional electrician may advise you to replace at least some of the components.

SEE ALSO: Supplementary bonding 10, Sockets for outdoor tools 58

Consumer units

• Important notes
Make a note of the last time you had your electrical installation checked professionally, and attach it as close as possible to the consumer unit. Similarly, jot down the next proposed date of inspection.

If your home is wired with both the old-style colour-coded cables and the newer ones, attach a note near the consumer unit to alert anyone working on the fixed wiring of your house in the future.

The consumer unit is the heart of your electrical installation: every circuit in your home has to pass through it. Although there are several different types and styles, all consumer units are based on similar principles.

Every consumer unit has a large main isolating switch, which can turn off the entire electrical system of the house. On some units, the switch is in the form of an RCD that can be operated manually but which will also 'trip' automatically should any serious fault occur, isolating the whole system in much less time than it would take for the electricity company's fuse to blow in a similar emergency. There is greater emphasis these days on protecting only the most vulnerable circuits with an RCD. With these 'split' consumer units, the main isolating switch still turns off every circuit simultaneously.

Some consumer units are designed in such a way that it's impossible to remove the outer cover without first turning off the main isolating switch. Even if yours is not this type, you should always switch off the power before exposing any of the elements within the consumer unit.

Having turned off the main switch, remove the cover (or covers) so that you can see how the unit is arranged. The cover (or covers) must be replaced before the unit is switched on again. Also, remember that even when the unit is switched off the cable connecting the meter to the main switch is still live – so take care.

Take note of the cables that feed the various circuits in the house. The blue-insulated neutral wires run to a common neutral block, where they are attached to their individual terminals. Similarly, the green-and-yellow earth wires run to a common earth block. The brown-covered live conductors are connected to terminals on individual fuseways or circuit breakers.

Some wires will be joined together in a single terminal. These are the two ends of a ring circuit, and that is how they should be wired.

Split consumer unit with miniature circuit breakers
Only the circuits on the left are protected by the RCD.

SEE ALSO: Colour coding 9, Main switch equipment 16, RCDs 17, Ring circuits 21

Fuses: types and ratings

In the consumer unit there is a fuseway for each circuit. Into the fuseway is plugged a fuse carrier, which is essentially a bridge between the main switch and that particular circuit. When the fuse carrier is removed from the consumer unit, the current cannot pass across the gap.

Identifying a fuse

Pull any of the fuse carriers out of the unit to see what kind of fuse it contains.

At each end of the carrier you will see a single-bladed or double-bladed contact. A rewirable carrier will have a thin wire running from one contact to the other, held by a screw terminal at each end. Fuse wire is available in various thicknesses, carefully calculated to melt at given currents when a circuit is substantially overloaded, thus breaking the 'bridge' and isolating the circuit.

Alternatively, the carrier may contain a cartridge fuse similar to those used in 13amp plugs, though circuit fuses are larger, varying in size according to their rating. The cartridge is a ceramic tube containing a fuse wire packed in fine sand. The wire is connected to metal caps at the ends of the cartridge that snap into spring clips on the contacts of the fuse carrier. Cartridge fuses provide better protection, since they blow faster than ordinary fuse wire; it is therefore advisable to use cartridge-fuse carriers wherever possible.

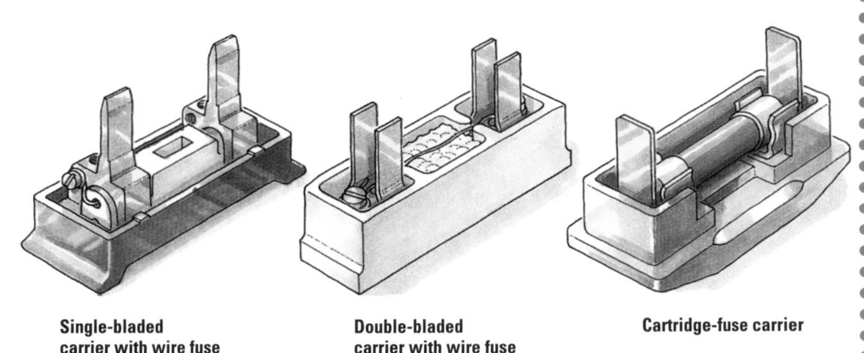

Single-bladed carrier with wire fuse

Double-bladed carrier with wire fuse

Cartridge-fuse carrier

Fuse ratings

Whatever the type of fuses used in the consumer unit, they are rated in the same way. Cartridge fuses are colour-coded and marked with the appropriate amp rating for a certain type of circuit. Fuse wire is bought wrapped round a card which is clearly labelled.

Never insert fuse wire that is heavier than the gauge intended for the circuit. To do so could result in a dangerous fault going unnoticed because the fuse wire fails to melt. And it is even more dangerous to substitute any other type of wire or metal strip; these provide no protection at all.

When you need to change a fuse, do not automatically replace it with one of the same rating. Check first that it is the correct type of fuse for the circuit. The fuse carrier should be marked and/or colour-coded. You can also look at the list of circuits printed on the inside of the consumer-unit cover to identify the carriers and their required ratings.

Keep spare fuse wire or cartridge fuses in or close to the consumer unit.

FUSE RATINGS

Circuit	Fuse	Colour coding
Door bell	5amp	White
Lighting	5amp	White
Immersion heater	15amp	Blue
Storage heater	15amp	Blue
Radial circuits – 20sq m maximum floor area 50sq m maximum floor area	20amp 30amp	Yellow Red
Ring circuits – 100sq m maximum floor area	30amp	Red
Shower unit	45amp	Green
Cooker	30amp	Red

MINIATURE CIRCUIT BREAKERS

Instead of fuses, miniature circuit breakers (MCBs) are often used to protect circuits. There are many types of MCB on the market, but only buy ones that are made to the required standards of construction and safety.

Make sure any MCB that you use is marked BS EN 60898, which is the relevant British Standard. There are also different classes of MCB (you need to look for Type B). And lastly, MCBs are classified according to the largest potential fault current they are able to clear; ask for M6 or M9, as these will clear any potential current likely to be met in a domestic situation. If for any reason these MCBs are unavailable, ask your electricity company whether they will accept alternatives.

MCB ratings

To conform to European standards, MCB ratings tend to vary slightly from circuit-fuse ratings. (See CIRCUITS: MAXIMUM LENGTHS.) However, it is perfectly acceptable if you have MCBs that match the slightly smaller ratings shown for circuit fuses.

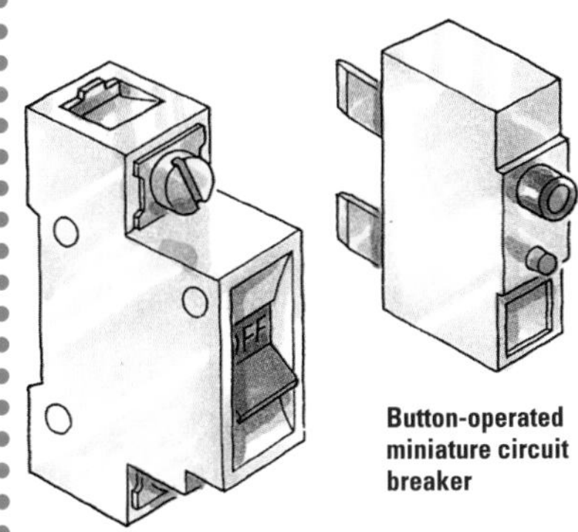

Switch-operated miniature circuit breaker

Button-operated miniature circuit breaker

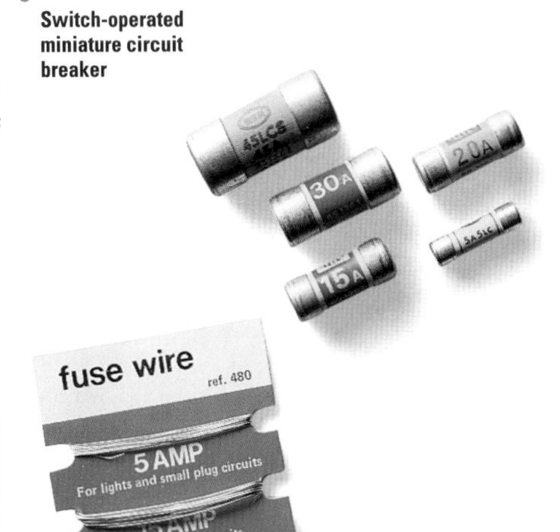

A selection of fuse wire and circuit fuses
From left to right: fuse wire, 45amp fuse, 30amp fuse, 20amp fuse, 15amp fuse, 5amp fuse.

Changing a fuse

When everything on a circuit stops working, first of all check the fuse to see if it has blown. Turn off the main switch on the consumer unit, take off the cover, and look for the failed fuse. To identify the relevant fuse, look at the list of circuits inside the cover. If there is no list, inspect the most likely circuits. If, for example, the lights blew when you switched them on, you need check only the lighting circuits, which are usually colour-coded white.

Checking a cartridge fuse

The simplest way to check a suspect cartridge fuse is to replace it with a new one and see if the circuit works.

Alternatively, you can check the fuse with a metal-cased torch. Remove the bottom cap of the torch, and touch one end of the fuse to the base of the battery while resting its other end against the torch's metal casing. If the torch bulb lights up, the fuse is sound.

Using a continuity tester
You can check a suspect cartridge fuse with a continuity tester. Place one of the tester's probes on each of the fuse's metal caps, then press the appropriate circuit-test button. If the indicator of the tester doesn't illuminate, the fuse has blown.

Testing a cartridge fuse
With the torch switched on, hold the fuse against the battery and the metal casing.

Checking a rewirable fuse

On a blown rewirable fuse, a visual check will usually detect the broken wire, plus scorch marks on the fuse carrier. If you cannot see the whole length of the fuse wire, pull gently on each end of the wire with the tip of a small screwdriver to see if it's intact.

Pull the wire gently with a small screwdriver

Replacing fuse wire

To replace blown fuse wire, loosen the two terminals holding the old wire and extract the broken pieces. Wrap one end of a new length of the correct type of fuse wire clockwise round one of the terminals and tighten the screw **(1)**. Then run the wire across to the other terminal, leaving it slightly slack, and attach it in the same way **(2)**. Cut off any excess wire from the ends.

If the wire passes through a tube in the fuse carrier, it has to be inserted before either terminal is tightened **(3)**.

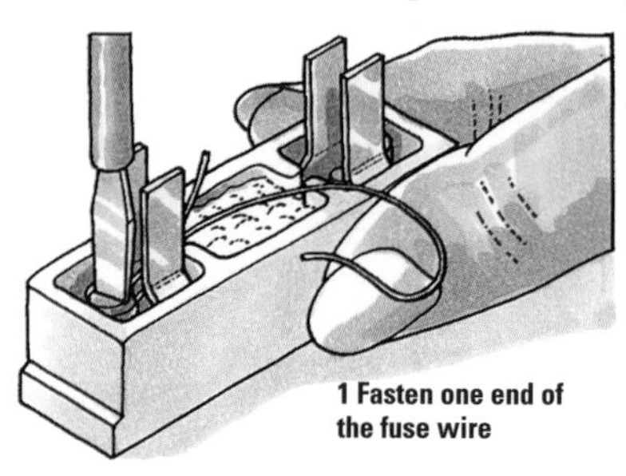

1 Fasten one end of the fuse wire

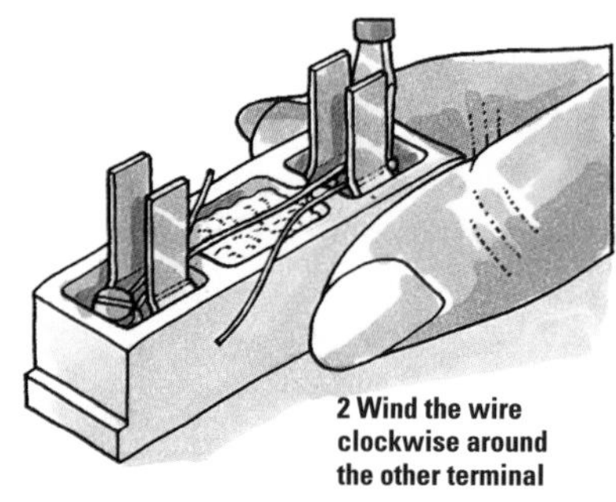

2 Wind the wire clockwise around the other terminal

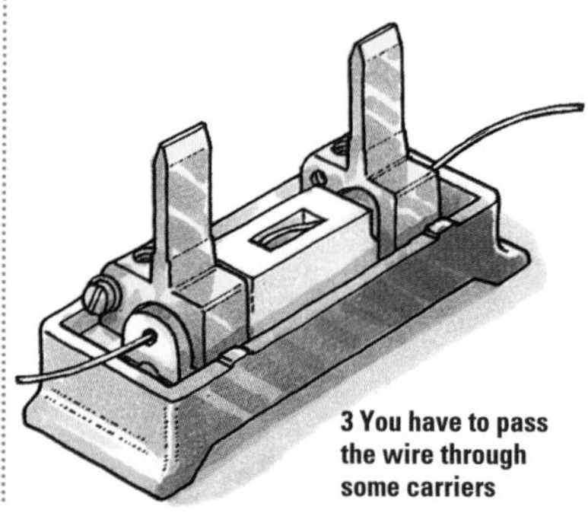

3 You have to pass the wire through some carriers

IF THE FUSE BLOWS AGAIN

If a replaced fuse blows again as soon as the power is switched on, then there is either a fault or an overload (too many appliances plugged in) on that circuit – and it must be detected and rectified before another fuse is inserted (see top right).

CHECKING OUT A FAULT

An electrician will test a circuit for you with special equipment – but first carry out some simple tests yourself.

Unplug all appliances on the faulty circuit to make sure that it is not simply overloaded, then switch on again.

If the circuit is still faulty, turn off the main switch on the consumer unit and, before inspecting any part of the circuit, remove the relevant fuse carrier or remove (or lock off) the MCB – so that no one can replace it while you are working. Inspect the relevant socket outlets and light fittings to see if a conductor has worked loose and is touching one of the other wires, or the terminals or outer casing, causing a short circuit.

If none of this enables you to find the fault, call in an electrician.

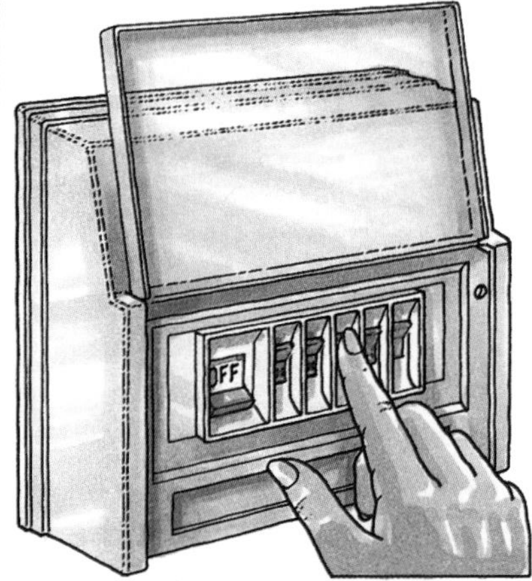

With the main switch off, reset the MCB

Resetting MCBs

In many consumer units you will find miniature circuit breakers (MCBs) instead of fuse carriers. Their current ratings tend to differ very slightly from fuse ratings, but the main difference is that circuit breakers switch to the 'off' position automatically, so a faulty circuit is obvious as soon as you inspect the consumer unit.

Turn the consumer unit's main switch off, and then simply close the switch on the miniature circuit breaker to reset it (there is no fuse to replace). If the MCB's switch or button won't stay in the 'on' position when power is restored, then there is still a fault on the circuit – which must be rectified.

Regular testing
It is a good idea to check that all your MCBs and RCDs are in good working order by tripping them deliberately every three months or so.

SEE ALSO: Consumer units 18, Fuses and fuse ratings 19, Circuit breakers 19

Domestic circuits

Running from the consumer unit are the cables that supply the various fixed wiring circuits in your home. Not only are the sizes of the cables different, the circuits themselves also differ, depending on what they are used for and also, in some cases, how old they happen to be.

Ring circuits

The most common form of 'power' circuit for feeding socket outlets is the ring circuit, or 'ring main'. With this method of wiring, a cable starts from terminals in the consumer unit and goes round the house, connecting socket to socket and arriving back at the same terminals. This means that power can reach any of the socket outlets or fused connection units from both directions, which reduces the load on the cable.

Ring mains are always run in $2.5mm^2$ cable and are protected by 30amp fuses or 32amp MCBs. Theoretically there is no limit to the number of socket outlets or fused connection units that can be fitted to a ring circuit provided that it does not serve a floor area of more than 100sq m (120sq yd) – a limit based on the number of heaters that would be adequate to warm that space. However, in practice two-storey houses usually have one ring main for the upper floor and another one for downstairs.

Spurs

The number of sockets on a ring main can be increased by adding extensions or 'spurs'. A spur can be either a single $2.5mm^2$ cable connected to the terminals of an existing socket or fused connection unit, or it can run from a junction box inserted in the ring.

Each (unfused) spur can feed only one fused connection unit for a fixed appliance or one single or double socket outlet. You can have as many spurs on a ring circuit as there were sockets on it originally (note that for this calculation a double socket is counted as two).

The 30amp fuse that protects the ring main remains unchanged, no matter how many spurs are connected to the circuit.

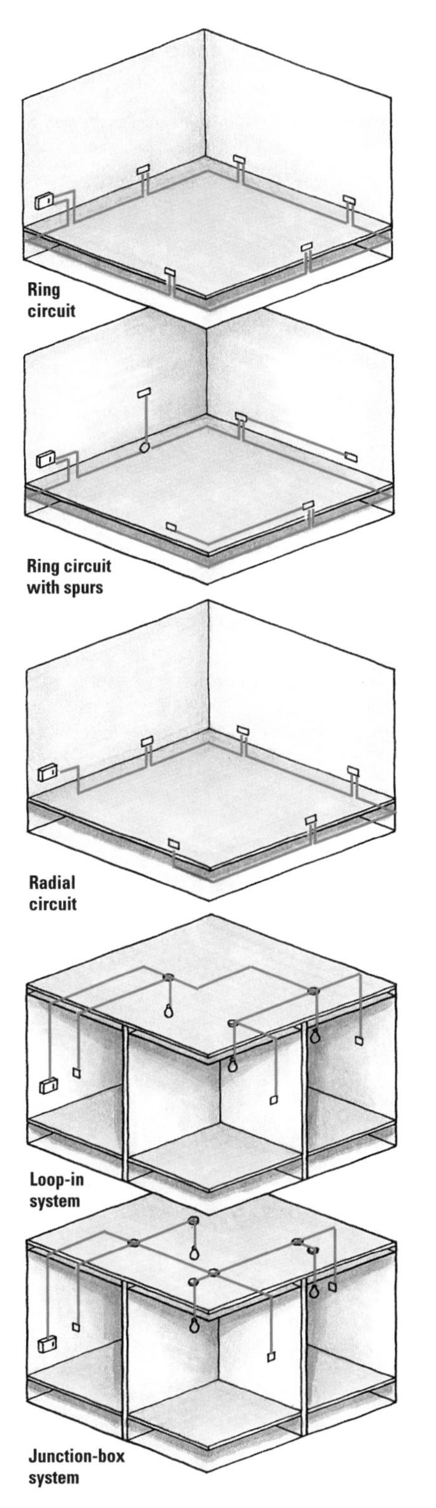

Radial circuits

A radial power circuit feeds a number of sockets or fused connection units – but, unlike a ring circuit, its cable terminates at the last outlet. The size of cable and the fuse rating depend on the size of the floor area to be supplied by the circuit. In an area of up to 20sq m (24sq yd), the cable needs to be $2.5mm^2$, protected by a 20amp MCB or a 20amp fuse of any type. For a larger area, up to 50sq m (60sq yd), you should use $4mm^2$ cable with a 30amp cartridge fuse or 32amp MCB (a rewirable fuse is not permitted).

Any number of socket outlets can be supplied by one of these circuits, and spurs can be added if required. These circuits are known as multi-outlet radial circuits. A powerful appliance such as a cooker or shower unit must have its own radial circuit.

Lighting circuits

Domestic lighting circuits are of the radial kind, but there are two systems currently in use.

The loop-in system simply has a single cable that runs from ceiling rose to ceiling rose, terminating at the last one on the circuit. Single cables also run from the ceiling roses to the various light switches.

The junction-box system (which is the older of the two systems) incorporates a junction box for each light. The boxes are situated conveniently on the single supply cable. A cable runs from each junction box to the ceiling rose, and another from the box to the light switch. In practice, most lighting circuits are a combination of the two methods.

A single circuit of $1mm^2$ cable is able to serve the equivalent of eleven 100W light fittings. Check the load by adding together the wattage of all the light bulbs on the circuit. If the total comes to more than 1200W, the circuit should be split. In any case, it makes sense to have two or more separate lighting circuits running from the consumer unit. If your house is large, requiring very long cable runs, have $1.5mm^2$ two-core-and-earth cable installed instead of $1mm^2$.

Lighting circuits must be protected by 5amp fuses or 6amp MCBs.

SEE ALSO: Fuse ratings 19, Cables 22, Socket outlets 28, Fused connection units 34, Cooker circuit 37, Shower circuit 47, Circuit lengths 66

Types of cable

Light-switch cabling
Wall-mounted light switches are usually wired with ordinary $1mm^2$ or $1.5mm^2$ two-core-and-earth cable. The live wire is colour-coded brown, and the switch-return wire blue. Because the blue wire carries live current back to the light fitting, it is normal to either attach a brown flag or slip brown sheathing over the wire. However, you can now buy special cable with two brown wires for wiring light switches.

Two-core-and-earth cable

Cable for the fixed wiring of electrical systems normally has three conductors: the insulated live and neutral ones and the earth conductor lying between them, which is uninsulated except for the sheathing that encloses all three conductors. Cable up to $2.5mm^2$ has solid single-core conductors; but larger sizes (up to $10mm^2$) wouldn't be flexible enough if they had solid conductors, so each one is made up of seven strands. The live conductor is insulated with brown PVC, and the neutral one with blue. If an earth conductor is exposed, as in a socket outlet, it should be covered with a green-and-yellow sleeve. Buy sleeving from any electricians' supplier.

Heat-resistant sleeving is available for covering the conductors in an enclosed light fitting, where the temperature could adversely affect the normal PVC insulation.

The PVC sheathing on the outside of the cable is usually white or grey.

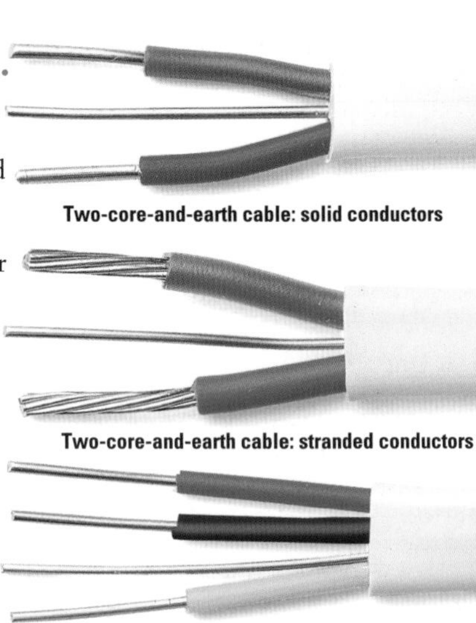

Two-core-and-earth cable: solid conductors

Two-core-and-earth cable: stranded conductors

Three-core-and-earth cable

This type of cable is used for a two-way lighting system, which can be turned on and off at different switches – at the top and bottom of a staircase, for example, so that you never have to use the stairs in the dark. It contains three insulated conductors – with brown, black and grey coverings – and a bare earth wire.

Three-core-and-earth cable

Single-core cable

Insulated single-core cable is used in buildings where the electrical wiring is run in metal or plastic conduit – a type of installation rarely found in domestic buildings. The cable is colour-coded in the normal way: brown for live, blue for neutral, and green-and-yellow for earth.

Single-core $16mm^2$ cable insulated in a green-and-yellow PVC covering is used for connecting the consumer unit to the earth. Single-core cable of the same size is used for connecting the consumer unit to the meter. The meter leads are insulated in brown for the live conductor and blue for the neutral one. The sheathing is grey.

Insulated single-core cable

Insulated-and-sheathed single-core cable

● **Cable sizes**
The chart on the right gives the basic sizes of cables used for wiring domestic circuits. For details of the maximum permitted lengths for circuits, see CIRCUITS: MAXIMUM LENGTHS.

If the company fuse is larger than 60amps, $25mm^2$ meter leads are required – but consult your local electricity company for advice.

CIRCUIT-CABLE SIZES

Circuit	Size	Type
Fixed lighting	$1.0mm^2$ & $1.5mm^2$	Two-core-and-earth
Bell or chime transformer	$1.0mm^2$	Two-core-and-earth
Immersion heater	$2.5mm^2$	Two-core-and-earth
Storage heater	$2.5mm^2$ & $4.0mm^2$	Two-core-and-earth
Ring circuit	$2.5mm^2$	Two-core-and-earth
Spurs	$2.5mm^2$	Two-core-and-earth
Radial – 20amp	$2.5mm^2$	Two-core-and-earth
Radial – 30amp	$4.0mm^2$	Two-core-and-earth
Shower unit	$10.0mm^2$	Two-core-and-earth
Cooker	$4.0mm^2$ & $6.0mm^2$	Two-core-and-earth
Consumer earth cable	$16.0mm^2$	Single core
Meter leads	$16.0mm^2$	Single core

Old cable may be dangerous and needs to be replaced

PREWAR CABLE

Houses that were wired before World War II may still have old cable that is sheathed and insulated in rubber, and some of them may even have old cable sheathed in lead.

Rubber sheathing is usually a matt black. It is more flexible than modern PVC insulation – unless it has deteriorated, in which case it will be crumbly.

Stripping cable

When cable is wired to an accessory, some of the sheathing and insulation must be removed. Slit the sheathing lengthwise with a sharp knife, peel it off the conductors, then fold it over the blade and cut it off. Take about 12mm (½in) of insulation off the ends of the conductors, using wire strippers.

Cover the uninsulated earth wire with a green-and-yellow plastic sleeve, leaving 12mm (½in) of the wire exposed for connecting to the earth terminal.

If more than one stranded conductor is to be inserted in the same terminal, twist the exposed ends together with strong pliers to ensure the maximum contact for all of the wires. Don't twist solid conductors; simply insert them together into the terminals and tighten the fixing screw. Pull on each conductor to make sure it is held securely.

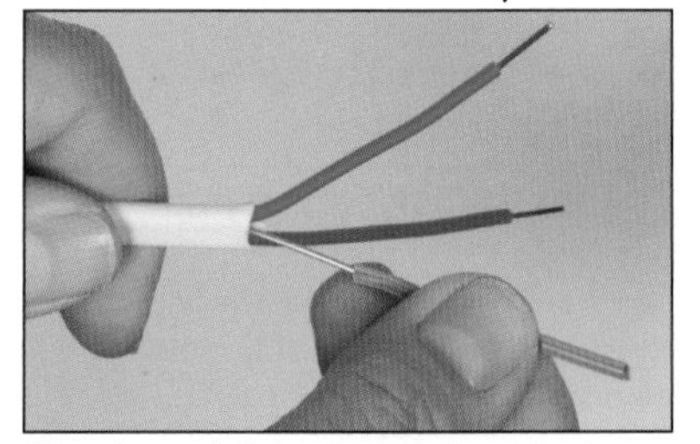

Slip colour-coded sleeving over the earth wire

SEE ALSO: Earth lead 16, Meter leads 16, Two-way lighting 54, Circuit lengths 66, Wire strippers 67

Running cable

To install a short cable run in a lath-and-plaster wall, hack the plaster away, fix the cable to the studs, and then plaster over again in the normal way.

Although you can run cable through the space between the two claddings of a partition wall, there is no way of doing this without some damage to the wall and the decoration. Drill a 12mm (½in) hole through the top wall plate above the spot where you are planning to position the switch, and then tap the wall directly below the hole to locate the nogging. Cut a hole in the lath-and-plaster to reveal the top of the nogging, then drill a similar hole through it.

Pass a weight on a plumb line through both holes, down to where the switch will be. Tie the cable to the line and pull it through.

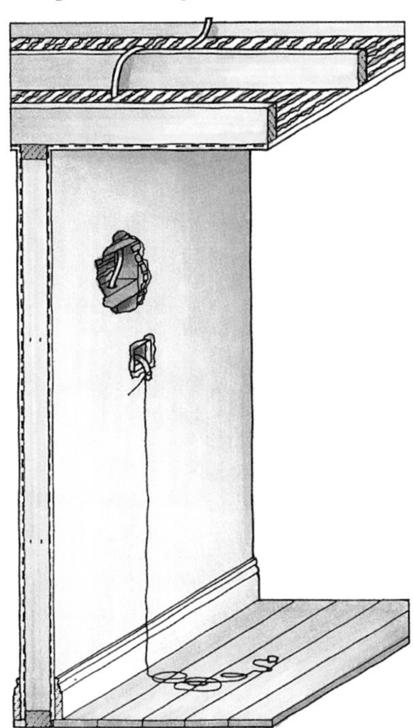

Running a cable through a hollow wall
If a nogging prevents you running cable directly to a switch, cut away some of the lath-and-plaster in order to drill a hole through the timber.

Long runs of cable are necessary to carry electricity from the consumer unit to all the sockets, light fittings and fixed appliances in the home. The cable must be fixed securely to the structure of the house along its route, except in confined spaces to which there is normally no access, such as inside hollow walls.

There are accepted ways of running and fixing cable, depending on particular circumstances.

Surface fixing

PVC-sheathed cable can be fixed to the surface of a wall or ceiling without any further protection. Fix it with plastic cable clips **(1)** or metal buckle clips **(2)** every 400mm (1ft 4in) on vertical runs, and every 250mm (10in) on horizontal runs. Try to keep the runs straight, and avoid kinks in the cable by keeping it on the drum as long as possible. If you do have to remove kinks, pull the cable round a thick dowel held in a vice.

If a cable seems vulnerable, or you simply want to hide it, run it inside plastic mini-trunking **(3)**. Screw or stick the trunking to the wall, then insert the cable and clip on the flexible cover strip.

1 Plastic cable clip

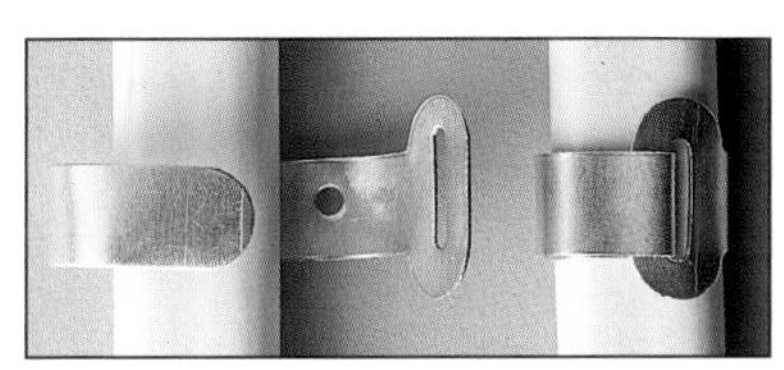

2 Metal buckle clip

Concealed fixing

While surface-fixed cable is acceptable in a cellar or in a garage or workshop, you wouldn't want to see it running across your living room walls or ceiling. From a decorative point of view, it's better to bury it in the plaster or hide it in a wall void. PVC-sheathed cable can be buried without further protection.

Where possible, run cable vertically to accessories such as switches or socket outlets, to avoid dangerous clashes with wall fixtures installed later. If that is not possible, you are permitted to run the cable horizontally directly from the switch or socket. However, if a cable isn't connected to a switch or socket on a wall in which it is concealed, then the cable must be within 150mm (6in) of the vertical or horizontal edges of the wall. Never, in any circumstances, run a buried cable diagonally across a wall.

Some people cover all buried cable with a plastic channel or run it inside conduit, but this is not required by the IEE Wiring Regulations. However, cable that is buried in plastic conduit can, if necessary, be withdrawn later without disturbing decorations.

Mark out your cable runs on the plaster, making allowance for a 'chase', or channel, about 25mm (1in) wide for single cable. Cut both sides with a bolster and club hammer, and then hack out the plaster between the cuts with a cold chisel. Normally, plaster is thick enough to conceal cable, but you may have to chop out some brickwork to get the depth. Clip the cable in the channel **(1)** and, once you have checked that the installation is working satisfactorily, plaster over it. To avoid electric shock, ensure that the power to that circuit is turned off before you use wet plaster round a switch or socket outlet **(2)**.

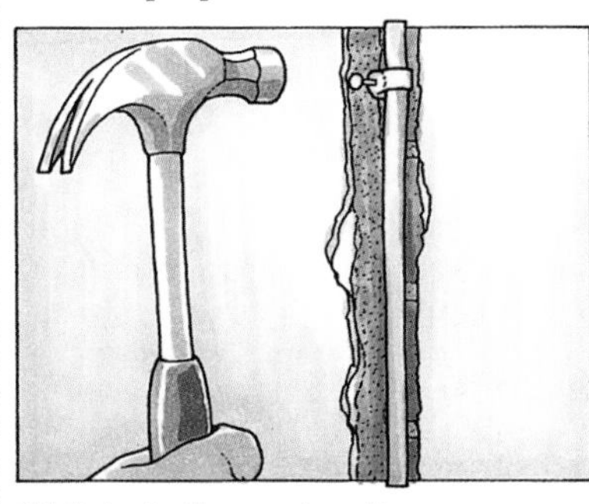

1 Nail plastic clips over the cable

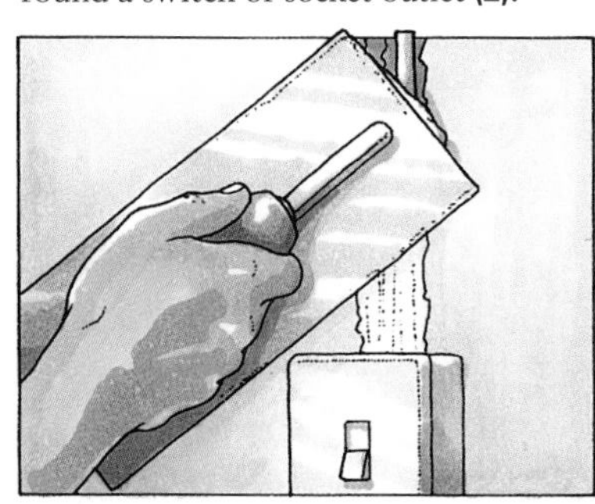

2 Repair the plaster up to the switch

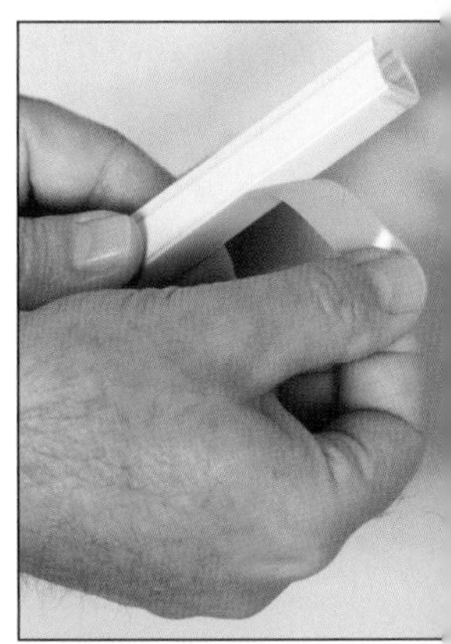

Peel off the backing

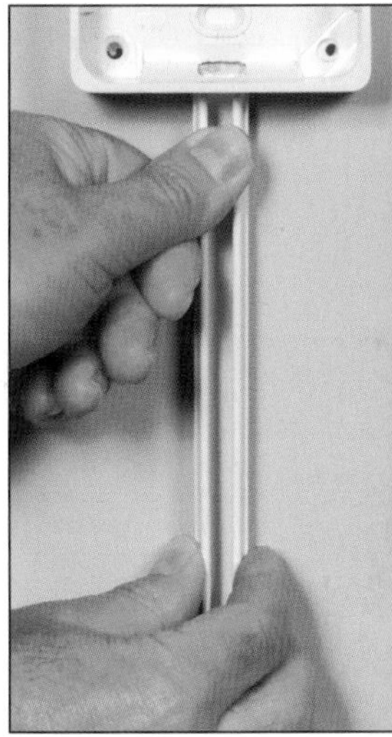

Press in position

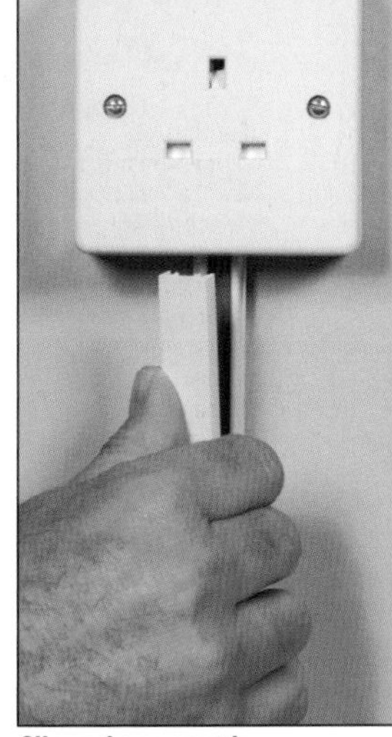

Clip on the cover strip

3 Using mini-trunking
Screw or stick mini-trunking to the wall and insert the cable. Make your connections, then snap on the cover strip.

SEE ALSO: Switching off 16

Running cable under floors

Power and lighting circuits are often concealed beneath floors. It isn't necessary to lift every floorboard to run a cable from one side of a room to the other: by lifting a board every 2m (6ft) or so, you should be able to pass the cable from one gap to the next with the help of a length of stiff wire bent into a hook at one end.

Look for floorboards that have been taken up before – they will be fairly easy to lift, so you will damage fewer boards.

Lifting floorboards

Lifting square-edged boards
Drive a wide bolster chisel between two boards about 50mm (2in) from the cut end of one of them **(1)**. Lever that board up with the bolster, then do the same on the other edge, working along the board until you have raised it far enough to wedge a cold chisel under it **(2)**. Proceed along the board, raising it with the chisel, till the board is loose.

Full-length boards
If you have to lift a board that runs the whole length of the floor from one skirting to the other, start somewhere near the middle of the board, close to one of the floor joists – the nail heads indicate the positions of joists. Lever the board up and make a sawcut across it, centred on the joist; then lift the board in the normal way.

Lifting tongue-and-groove boards
You cannot lift a tongue-and-groove floorboard until you have cut through the tongues along both sides of the board with a floorboard saw, which has a blade with a rounded tip.

Alternatively, use an electrician's 'skate' – which has a cutting disc that fits between the boards. Run the tool back and forth with one foot.

Cutting a full-length board
Saw a full-length board in two directly over a floor joist.

Using a skate
Run the disc of an electrician's skate between tongue-and-groove boards.

1 Prise up the floorboard with a bolster

2 Wedge the raised end with a cold chisel

Cutting a board next to a skirting

A joist that is fitted close to a wall may make it impossible to lift a floorboard in the normal way without damaging the bottom edge of the skirting.

In such a case, drill a starting hole through the floorboard alongside the joist, then insert the blade of a padsaw into the hole and cut across the board, flush with the side of the joist **(1)**.

To support the cut end afterwards, nail a length of 50 x 50mm (2 x 2in) softwood to the joist. Hold the batten tightly against the undersides of the adjacent floorboards while you are fixing it, to ensure that the cut board will lie flush with the others **(2)**.

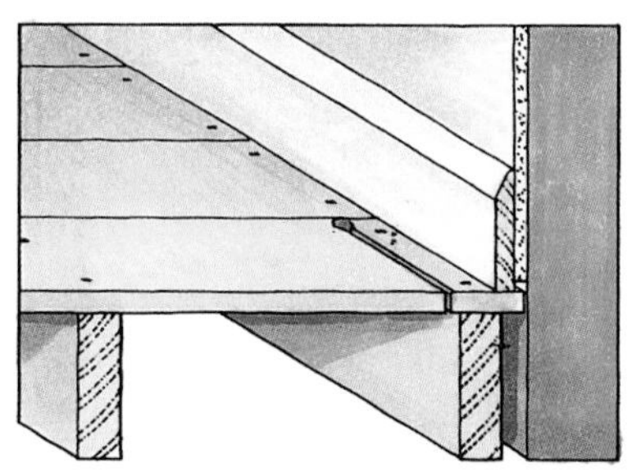

1 Cut through a trapped board with a padsaw

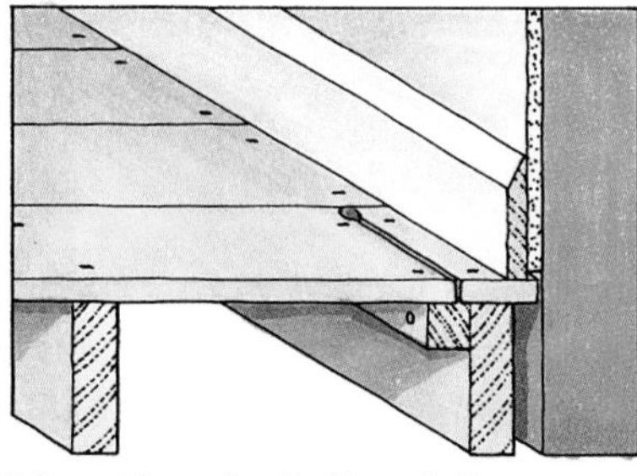

2 Support the cut board with a nailed batten

● **Underfloor wiring**
If you have to drill holes through the joists when running cable under a suspended floor, make sure you do not weaken the structure.

Solid floors

In a new concrete floor, you can lay conduit and then run cable through it before the concrete is poured.

In an existing solid floor, you can cut a channel for conduit, although it's hard work without an electric hammer and chisel bit; and if the floor is tiled, you will not want to spoil it for one or two socket outlets. An alternative is to drop spur cables, buried in the wall plaster, from the ring circuit in the upper floor.

Another way is to run cable through the wall from an adjacent area and channel it horizontally in the plaster just above the skirting. Yet another is to install the type of skirting-height plastic trunking that is designed to house cable and socket outlets. Often used in commercial premises, this kind of trunking may be difficult to obtain from smaller retailers.

In the roof space

All wiring can be surface-run in the roof space; but as people may enter the roof space from time to time, you must make sure the cable is clipped securely to the joists or rafters. Run it through holes in the normal way, especially where joists are to be boarded over or in areas of access – around water tanks and near the entrance hatch, for example. If short lengths have to run on top of a joist, add mechanical protection.

Wiring overlaid by roof-insulation material has a slightly higher chance of heating up. Lighting circuits do not present a problem; but circuits on which there are heaters, cookers or shower units, for example, are more critical. Wherever possible, run cable over thermal insulation. If you cannot avoid running it under the material, use a heavier cable – but consult a qualified electrician to be on the safe side.

When expanded-polystyrene insulation is in contact with electrical cable for a long time, it affects the plasticizer in the PVC sheathing on the cable. The plasticizer moves to the surface of the sheathing, reacts with the polystyrene, and forms a sticky substance on the cable. This becomes a dry crust which cracks if the cable is lifted out of the roof insulation and bent. Although it gives the impression that the cable insulation is cracking, scientific testing has shown that the cracking is in fact merely in the surface crust. On balance, however, it is best to keep cable away from polystyrene.

SEE ALSO: Mini-trunking 23, Spur cables 31

Running cable through the house

Running cable through the house structure
Use the most convenient method to run cable to sockets and switches.
1 Clip cable to battens nailed to roof timbers in the loft.
2 Junction boxes must be fixed securely.
3 In the joists near the hatch, run cable through holes.
4 Run cable over loft insulation.
5 To avoid damaging a finished floor, you can run a short spur through the wall from the next room.
6 When cable needs to run across the line of joists, drill holes 50mm (2in) below the joists' top edges.
7 When cable needs to run parallel to the joists, it can lie on the ceiling below.
8 Let cable drape onto the base below a suspended floor.
9 If it's impractical to run cable through a concrete floor, you can drop a spur from the floor above, but label the consumer unit accordingly.

● **Labelling circuits**
If you have added sockets to a ring or radial circuit, make sure that the label in the consumer unit clearly identifies the circuit to which the new sockets are connected.

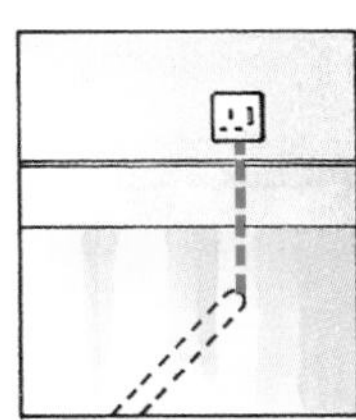

Burying cable in concrete
When you are laying a new concrete floor, take the opportunity to bury conduit for cable.

Running the cable

On the ground floor the cable can rest on the earth or on the concrete platform below the joists, provided that there won't normally be access to the space. Allow enough slack, so the cable isn't suspended above the platform, which might put a strain on fixings to junction boxes or socket outlets. For the same reason, beside junction boxes or other accessories, secure cable with clips to the side of the joist. Never attach circuit cable to gas or water pipes; and don't run it next to heating pipes, as the heat could melt the insulation.

When laying cable between a floor and the ceiling below, it can rest on the ceiling without any other fixing, so long as the cable runs parallel with the joists. If it runs at right angles to the joists, drill a series of 12mm (½in) holes, one through each joist along the intended cable run. The holes must be at least 50mm (2in) below the tops of the joists, so floorboard nails won't at some time be hammered through the cable. Similarly holes must be at least 50mm (2in) from the bottom edge of ceiling joists, in order to be certain that nails driven from below cannot pierce the cable. The space between joists is limited, but you can cut down a spade bit for use in a power drill.

Drilling the joists
Shorten a spade bit so that your drill fits between the joists. Take care not to weaken joists.

Having marked out the position of a socket or fused connection unit, cut a channel from it down to the skirting board and, with an extra-long masonry bit fitted in a power drill, remove the plaster from behind the skirting board. By using the drill at a shallow angle you can loosen much of the debris, but you will probably have to finish the job with a slim cold chisel. Raking the debris out from below with the same chisel also helps to dislodge it.

Drilling behind skirting
Use an extra-long masonry bit to remove plaster behind a skirting board.

Pass a length of stiff wire with one end formed into a hook down behind the skirting. Hook the cable and pull it through, at the same time feeding it from below with your other hand.

Fitting a blind grommet
There should be only just enough room for a cable to pass through a grommet into a mounting box for a switch or socket.

Preventing the spread of fire

Every time you cut an opening in the structure of the house for a cable, you are creating a potential route for fire and smoke to spread. After you have installed the cable, fill any holes between floors or rooms, using plaster or some other non-flammable material (not asbestos). Even where you pass a cable into a mounting box, you must fit a 'blind' grommet and cut a hole through it that is only just large enough for the cable.

SEE ALSO: Cable clips 23, Concealing cable 23, Fitting a grommet 29, Running a spur 31

Assessing your installation

Inspect your electrical system to ensure that it is safe and adequate for your future needs. But remember, you should never examine any part of it without first switching off the power at the consumer unit.

If you are in doubt about any aspect of the installation, don't hesitate to ask a qualified electrician for an opinion. If you get in touch with your local electricity company, they will arrange for someone to test the whole system for you. There is usually a charge for this service. It is recommended that you should have your house wiring tested at least every ten years.

QUESTIONS	ANSWERS
Do you have a modern consumer unit, or a mixture of old 'fuse boxes'?	Old fuse boxes can be unsafe and should be replaced with a modern unit. Seek professional advice about this.
Is the consumer unit in good condition?	Replace a broken casing or cracked covers. Check that all the fuse carriers are intact and that they fit snugly in the fuseways.
Are the fuse carriers for the circuits clearly labelled?	If you cannot identify the various circuits, have an electrician test the system and label the fuses.
Are all your circuit fuses of the correct ratings?	Replace any fuses of the wrong rating. If an unusually large fuse is protecting one of the circuits, don't change it without getting professional advice – it may have a special purpose. If you find any wire other than proper fuse wire in a fuse carrier, replace it at once.
Are the cables that lead from the consumer unit in good condition?	The cables should be fixed securely, with no bare wires showing. If the cables appear to be insulated with rubber, have the whole installation checked as soon as possible. Rubber insulation has a limited life, so yours could already be dangerous.
Is the earth connection from the consumer unit intact and in good condition?	If the connection seems loose or corroded, have the electricity company check whether the earthing is sound. You can check an RCD by pushing the test button to make sure it is working mechanically.
What is the condition of the fixed wiring between floors and in the loft or roof space?	If just a few cables appear to be rubber-insulated, have the entire system checked by a professional – it can be confusing, as old cable may have been disconnected but left in place during a previous upgrade. If cable is run in conduit, it can be difficult to check on its condition – but if it looks doubtful where it enters accessories, have the circuit checked professionally. Wiring should be fixed securely and sheathing should run into all accessories, with no bare wire in sight. Junction boxes on lighting circuits should be screwed firmly to the structure and should have their covers in place.
Is the wiring unobtrusive and orderly?	Tidy all surface-run wiring into straight properly clipped runs. Better still, bury the cable in the wall plaster or run it under floors and inside hollow walls.
Are there any old round-pin socket outlets?	Make sure their wiring is adequate. Replace old radial circuits with modern wiring and 13amp square-pin sockets as soon as possible.
Are the outer casings of all accessories in good condition and fixed securely to the structure?	Replace any cracked or broken components and secure any loose fittings.
Do switches on all accessories work smoothly and effectively?	If the switches are not working properly, replace the accessories.
Are all the conductors inside accessories connected securely to their terminals?	Tighten all loose terminals and ensure that no bare wires are visible. Fit green-and-yellow sleeves to earth wires if they have not been fitted.

SEE ALSO: Switching off 16, Earth connection 16–17, Old fuse boards 17, RCDs 17, Fuse ratings 19, Replacing fuses 20, Old cable 22, Running cable 23–5, Replacing sockets 30, Radial circuit 33, Replacing switches 52

Assessing your installation

QUESTIONS	ANSWERS
Is insulation around wires inside any accessories dry and crumbly?	If so, it is rubber insulation in advanced decomposition. Replace the covers carefully and have a professional check the system as soon as possible.
Do any sockets, switches or plugs feel warm? Is there a burning smell ? Or are there scorch marks visible on sockets or around the base of plug pins? Does a socket spark when you pull out a plug? Or a switch when you operate it?	These symptoms mean loose connections in the accessory or plug, or a poor connection between plug and socket. Tighten loose connections and clean all fuse clips, fuse caps and plug pins with silicon-carbide paper, then wipe them with a soft cloth. If the fault persists, try fitting a new plug. If that fails to cure the problem, replace the socket or switch.
Is it difficult to insert a plug in a socket?	The socket is worn and should be replaced.
Are your sockets in the right places?	Sockets should be placed conveniently round a room so that you need never have long flexes trailing across the floor or under carpets. Add sockets to the ring circuit by running spurs or by extending the circuit.
Do you have enough sockets?	If you have to use plug adaptors, you need more sockets. Replace singles with doubles, add spurs, or extend the ring circuit.
Is there old braided twin flex hanging from some ceiling roses?	Replace it with PVC-insulated-and-sheathed flex. Also check that the wiring inside the rose is PVC-insulated.
Are there earth wires inside your ceiling roses?	If not, get professional advice on whether to replace the lighting circuits.
Is your lighting efficient?	Consider extra sockets or different light fittings to make your lighting more effective. Make sure you have two-way switching on stairs.
Is there power in the garage or workshop?	Detached outbuildings should have their own power supply.

(from left to right:)

Scorch marks
Scorch marks on a socket or round the base of plug pins indicate poor connections.

Overloaded socket
If you have to use adaptors to power your appliances, you should fit extra sockets.

Unprotected connections
Sheathe any bare earth wires and make sure covers or faceplates are fitted to all accessories.

(from left to right:)

Incorrect fuse
Replace improper wire with fuse wire.

Round-pin socket
Replace old round-pin sockets with 13amp square-pin sockets.

Damaged socket
Replace cracked or broken faceplates.

SEE ALSO: Flex 12, Replacing fuses 20, Replacing sockets 30, Running a spur 31, Extending ring circuit 32, Replacing switches 52, Two-way switching 54, Wiring outbuildings 64–5

Socket outlets

Whatever type of circuits exist in your home, only use 13amp square-pin sockets. All round-pin sockets are now out of date – and even if they are not actually dangerous at the moment, you should have them checked and consider changing your wiring to accommodate 13amp sockets.

Before you start work on any socket, switch the power off at the consumer unit and remove the fuse or remove (or lock off) the MCB for the relevant circuit – then test the socket with an appliance you know to be working, to make sure the socket has been switched off properly.

Types of 13amp socket

Triple sockets
Triple sockets are useful where several electrical appliances are grouped together.

Although all sockets are functionally similar, there are several variations to choose from. For most situations, you are likely to use either a single or double socket. Both are available switched or unswitched, and with or without neon indicators so you can see at a glance whether the socket is switched on. All of these are wired in the same way.

Another basic difference is in how the sockets are mounted. They can either be surface-mounted (screwed to the wall in a plastic box) or flush-mounted in a metal box buried in the wall, with only its faceplate visible.

Switched single   Unswitched single

Switched double

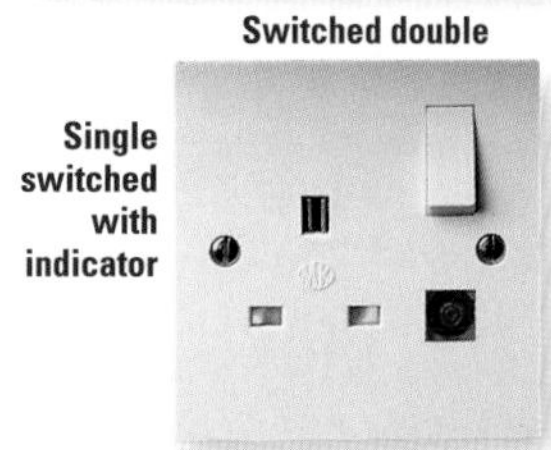

Single switched with indicator

Positioning socket outlets

Choose the most convenient positions for hi-fi and computer equipment, table lamps, television, and so on, and position your sockets accordingly. To avoid using adaptors or long leads, distribute the sockets evenly round living rooms and bedrooms, and wherever possible fit doubles rather than singles. Don't forget sockets for running a vacuum cleaner in hallways and on landings.

The recommended position for a socket is between 450 and 1200mm (1ft 6in to 4ft) above the floor, which makes it accessible to a person using a wheelchair. Although it is not a mandatory requirement, you may want to take this recommendation into account when rewiring an existing property.

In the kitchen, fit at least four double sockets 150mm (6in) above the worktops. You will also need sockets for floor-standing appliances such as refrigerators and dishwashers.

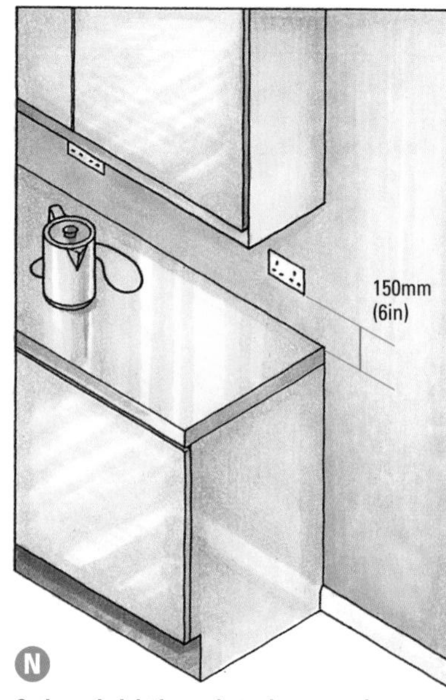

Optimum height for sockets above a worktop

Surface-mounting socket outlets

First, break out the thin plastic webs that cover the fixing holes in the back of a plastic mounting box. The best tool to use for this is an electrician's screwdriver. Two fixings should be sufficient. The fixing holes are slotted to enable easy adjustment.

Hold the mounting box firmly against a masonry wall – at the same time levelling it with a small spirit level – and mark the position of the fixing holes on the wall with a bradawl through the holes in the back of the box. Drill and plug the holes with No 8 wallplugs.

With a larger screwdriver and pliers, break out the plastic web covering the most convenient cable-entry hole in the box. For surface-run cable this will be in the side; for buried cable it will be the one in the base.

Feed the cable into the mounting box to form a loop about 75mm (3in) long **(1)**, then fix the box to the wall with 32mm (1¼in) countersunk woodscrews. Finally, wire and fit the socket.

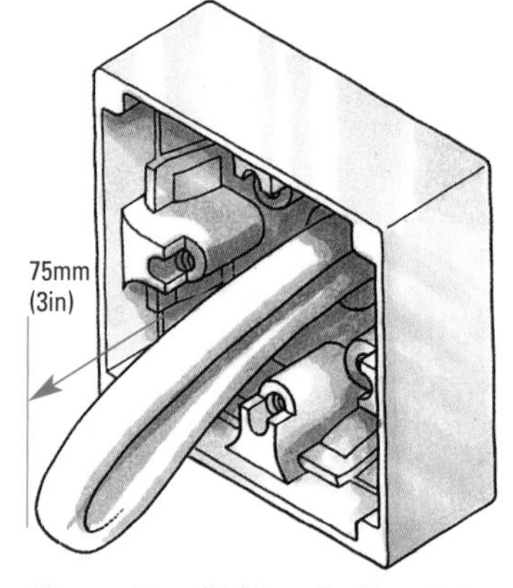

1 Leave a 75mm (3in) loop of cable at the box

Fixing to a hollow wall

On a dry-partition or lath-and-plaster wall, a surface-mounted box is fixed with any of the standard fixings used for hollow walls. Alternatively, use ordinary woodscrews if you are able to position the box over a stud – in which case, make sure you can feed the cable into the mounting box past the stud **(2)**.

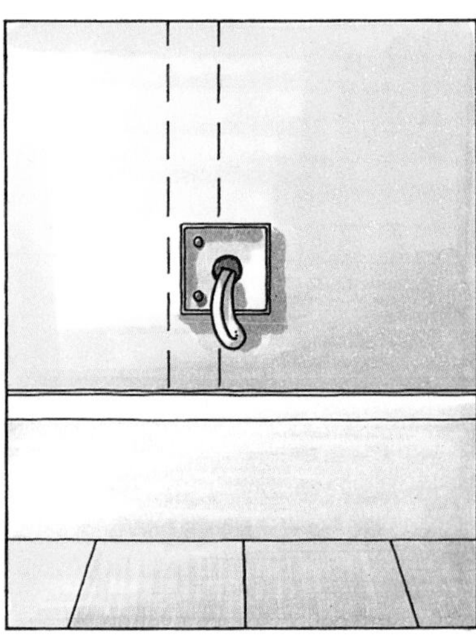

2 Feed the cable into the box past the stud

SEE ALSO: Switching off 9, 16, Wiring a socket 30, Wiring kitchen appliances 36, Circuit lengths 66

Flush-mounted sockets

Fixing to masonry

Hold the metal box against the wall and draw round it with a pencil **(1)**, then mark a 'chase' (channel) running up from the skirting to the box's outline.

Using a bolster or cold chisel, cut away the plaster, down to the brickwork **(2)**, within the marked area.

With a masonry bit, bore several rows of holes down to the required depth **(3)** across the recess for the box; then, using a cold chisel, cut away the brick to the depth of the holes, so that the box will lie flush with the plaster.

Try the box in the recess. If it fits in snugly, mark the wall through the fixing holes in its back, then drill the wall for screw plugs. If you have made the recess too deep or the box rocks from side to side, apply some filler in the recess and press the box into it, flush with the wall and properly positioned. After about 10 minutes, ease the box out carefully and leave the filler to harden, so that you can mark, drill and plug the fixing holes through it.

Next, knock out one or more of the blanked-off holes in the box to accommodate the cable. Fit a blind grommet into each hole to protect the cable's sheathing from the metal edges **(4)**, feed the cable into the box, and screw the box to the wall.

Plaster up to the box and over the cable chased into the wall; then, when the plaster has hardened, wire and fit the socket itself.

Fixing to plasterboard

In order to fit a flush socket to a wall made of plasterboard laid over wooden studs, trace the outline of the metal box in position on the wall and drill a hole in each corner of the outline. Then use a padsaw to cut out the recess for the box.

After punching out the blanked-off entry holes in the box and fitting rubber grommets, feed the cable into the box.

Clip dry-wall fixing flanges to the sides of the box **(5)**. These will hold it in place by gripping the wall from inside. Ease one side of the box, with flange, into the recess; and then, holding the screw-fixing lugs in order not to lose the box, manoeuvre it until both flanges are behind the plasterboard and the box sits snugly in the hole. (See also far right.)

Finally, wire and fit the socket. As you tighten up the fixing screws, the plasterboard will be gripped between the flanges and the faceplate.

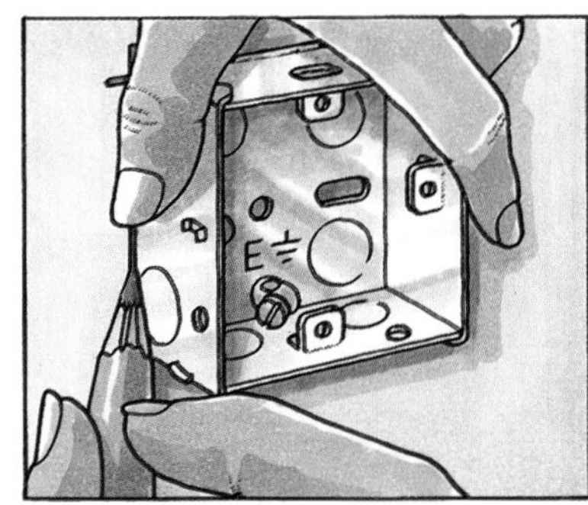

1 Draw round the mounting box

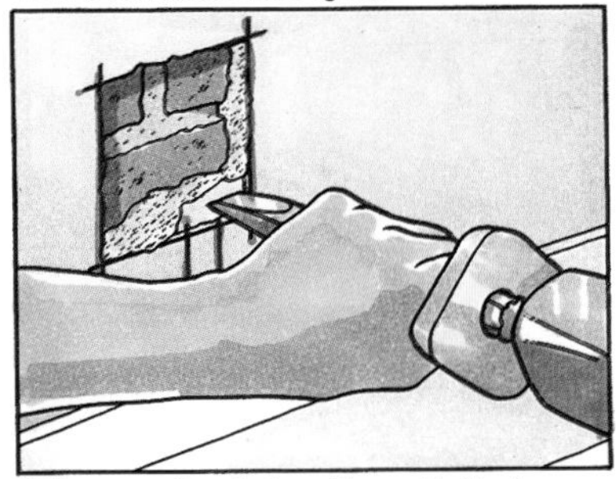
2 Chop away the plaster with a cold chisel

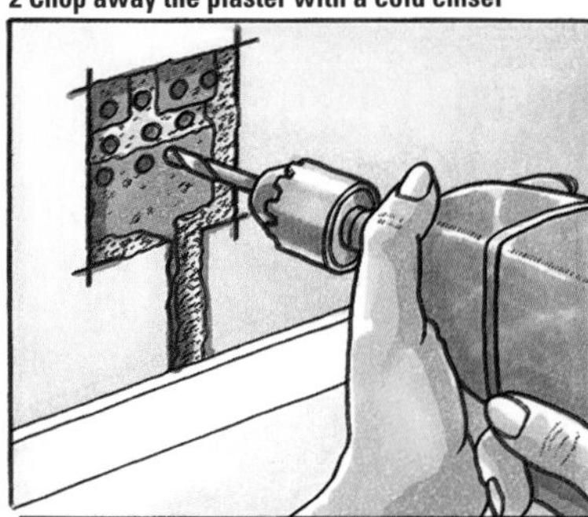
3 Drill out the brickwork with a masonry bit

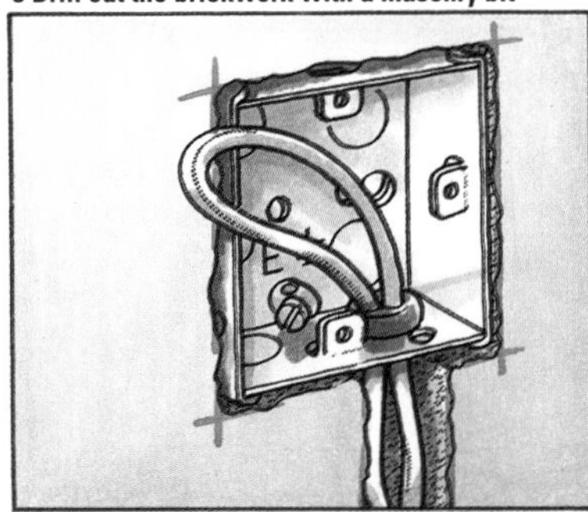
4 Fit a soft grommet in the cable-entry hole

5 Dry-wall fixing flanges clipped to a box

FLUSH-MOUNTING TO LATH-AND-PLASTER

If you want to fit a flush socket outlet in a lath-and-plaster wall, try to locate it over a stud or nogging.

Mark the position of the metal box, cut out the plaster, and saw away the laths with a padsaw. Try the box for fit, and if necessary chop a notch in the woodwork until the box lies flush with the wall surface **(1)**. Feed in the cable; and screw the box to the stud before wiring and fitting the socket.

If you can't position the socket on a stud, cut away enough of the plaster and laths to make a slot in the wall running from one stud to the next. Between the studs, screw or skew-nail a softwood nogging to which you can fix the box. If need be, set the batten back from the front edges of the studs, to make the box lie flush with the wall surface **(2)**. Feed the cable into the box and make good the surrounding plaster before you wire and fit the socket.

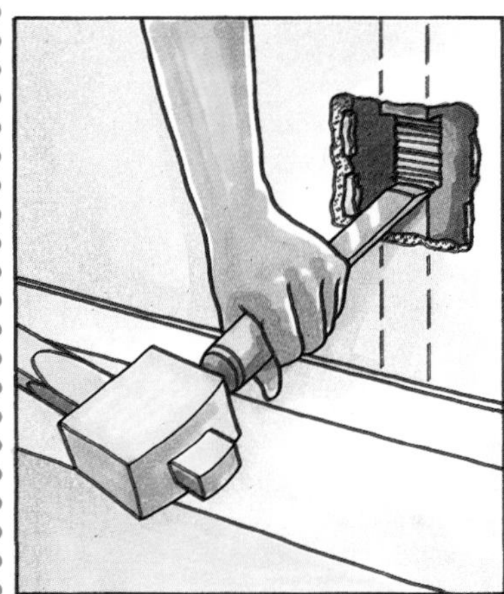
1 Notch a wall stud for a mounting box

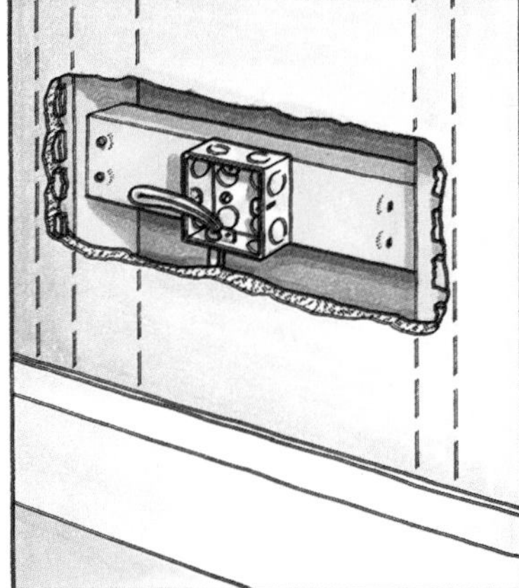
2 Nail a nogging between studs
Cut away wall plaster and laths when you have to fix a mounting box between wall studs.

Cavity-wall box
Instead of fitting dry-wall fixing flanges to a standard mounting box, you can use a special cavity-wall box with integral hinged flanges that you push through the sides of the box after it is fitted.

Replacing socket outlets

If you need to replace a broken or faulty socket, there are several options worth considering before you embark on the job.

Simple replacement

Replacing a damaged socket with a similar one is a fairly straightforward job. A socket outlet of any style will fit into a metal mounting box, but check carefully when you substitute a socket that screws to a surface-mounted plastic box. Although it will fit and function perfectly well, square corners and edges on either will not suit rounded ones on the other – in which case, you may also have to buy a new, matching box.

An unswitched socket outlet can be replaced with a switched one without any change to the wiring or fixing.

Switch off the power at the consumer unit and take out the circuit fuse, then remove the fixing screws from the faceplate and pull the socket out of the box.

Loosen the terminals to free the conductors. Check that all is well inside the box, then connect the conductors to the terminals of the new socket. Fit the faceplate, using the original screws if those supplied with the new socket don't match the thread in the box.

Surface to flush

If you have to renew a surface-mounted socket for any reason, you may want to take the opportunity to replace it with a flush one.

Turn off the power, remove the old socket and box, and then recess the new metal box into the wall, taking care not to damage the cable.

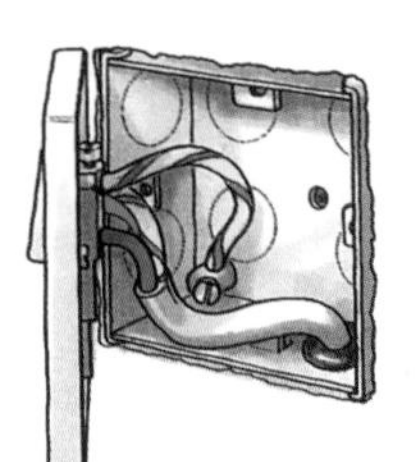

Flying-earth leads
Short lengths of cable are sometimes found running from the earth terminal on the socket outlet to a terminal inside a metal mounting box. This does no harm and the leads can be left in place – but it is not necessary to provide them on your new wiring, provided the metal box has at least one fixed lug for attaching the faceplate. However, flying earths are necessary if the earth connection is provided by a metal conduit or sheath system.

Replacing a single socket with a double

One way to increase the number of socket outlets in a room is to substitute doubles for singles. Any single socket on a ring circuit can be replaced with a double without making any changes to the wiring.

A single socket on a spur can be replaced with a double one so long as it's the only socket on that spur – it needs to be connected to a single cable. To ensure that a socket fed by two cables is not one of two sockets on the same spur (which is no longer permitted), carry out the ring-circuit continuity test – see opposite.

Remember to switch off the power before making any alterations.

Surface to surface

Replacing a surface-mounted single socket with a surface-mounted double is easy. Having removed the old socket outlet, simply fix the new, double box to the wall in the same place.

Flush to surface

To avoid the disturbance to decor that is involved in installing a flush double socket, fit a socket converter, which is made with two fixing holes that will line up with the fixing lugs on the buried metal box **(1)**. Although it isn't as slim as a standard flush-mounted socket, the faceplate of a socket converter is only 20mm (¾in) thick. Wire the existing cable into the back of the converter.

Alternatively, you can fit an ordinary double surface-mounted box to the fixing lugs of the buried metal box, and fit a standard socket outlet.

Flush to flush

Remove the old single socket and its metal box, then try the new double box over the hole. You can either centre the box over the hole or align it with one end **(2)**, whichever is more convenient. Trace the outline of the box on the wall and cut out the brickwork.

Use a similar procedure to substitute a double socket for a single in a hollow wall, installing the socket by whichever method is most convenient.

Surface to flush

To replace a single surface-mounted socket with a flush double, cut a recess for the metal box in the normal way.

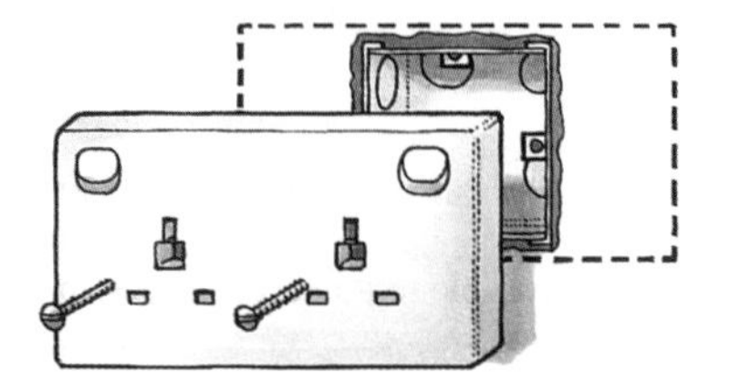

1 Fixing a socket converter over a flush box

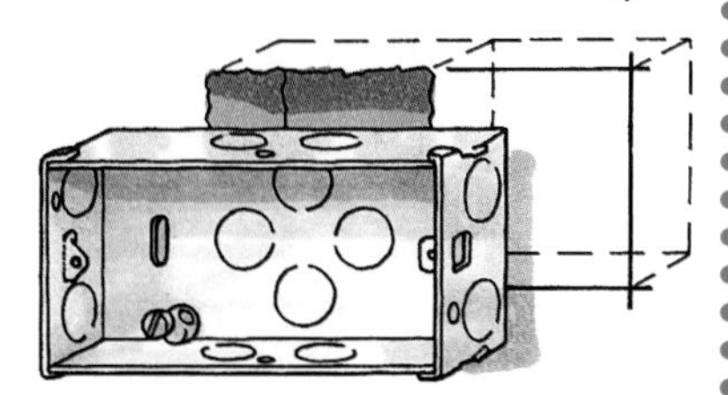

2 Cut out extra brickwork for a double box

WIRING A SOCKET OUTLET

When a single cable is involved, strip off the sheathing in the normal way and connect the wires to the terminals: the blue wire to neutral – N; the brown one to live – L; and the earth wire, which you should insulate yourself with a sleeve, to earth – E **(1)**. If necessary, fold the stripped ends over, so that no bare wire protrudes from a terminal.

Connecting to a ring-circuit cable

When connecting to a ring circuit, cut through the loop of cable and strip the sheathing from each half. Insert the bared ends of matching wires – live with live and so on – into the terminals **(2)**. Slip sleeves onto the earth wires. After tightening the terminal screws, pull on each wire to ensure it is fixed securely.

Cable is stiff, which can make it difficult to close the socket faceplate, so bend each conductor until it folds into the mounting box. Locate both of the fixing screws and tighten them gradually in turn until the plate fits firmly in place against the wall or box.

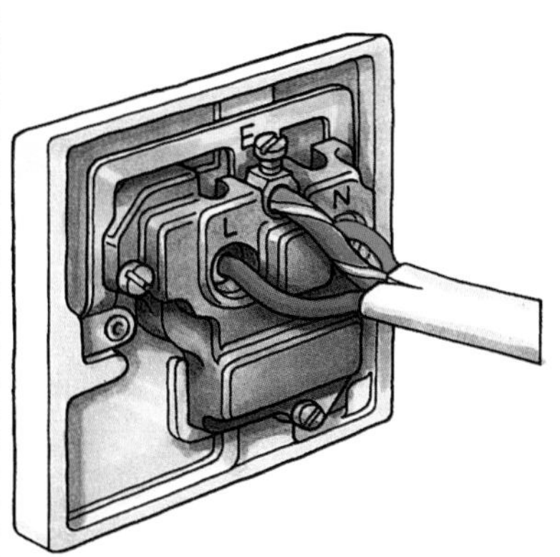

1 Wiring a socket outlet

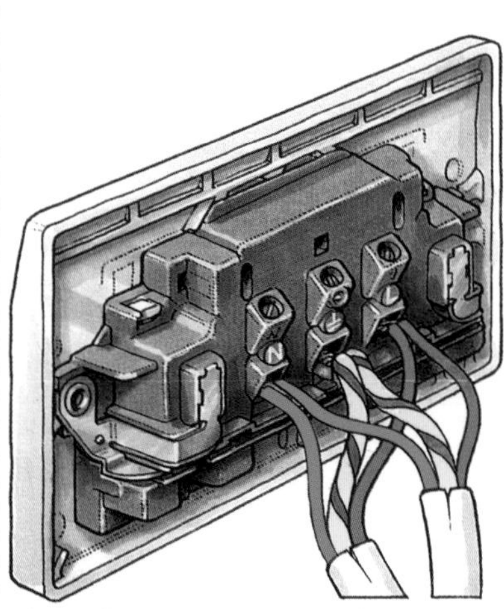

2 Connecting a socket to a ring circuit

SEE ALSO: Switching off 16, Stripping cable 22, Types of socket 28, Mounting to a hollow wall 29, Recessing a metal box 29

Adding a spur to a ring

If you need more sockets in a room, you can run 2.5mm² spur cables from a ring circuit and have as many spurs as there are sockets already on the ring. A spur can feed one single or one double socket.

A spur cable can be connected to any socket or fused connection unit on the ring circuit, or to a new junction box inserted in the circuit. If running a spur cable from an existing socket would mean disturbing the plaster, it will be more convenient to use a junction box; and if there is no socket outlet within easy reach of the proposed new one, using a junction box may save cable. If the cable is surface-run and you want to extend a row of sockets – behind a workbench, for example – then it will be simpler to connect the spur to a socket.

Examine the socket. If it is fed by a single cable, it is probably already on a spur; and if there are three cables in the socket, then it's already feeding a spur itself. What you need to look for is a socket that has two cables – but before you connect the spur to it, carry out a continuity test to make sure the socket is actually on a ring.

Testing for continuity

Isolate the ring circuit by switching off, then lock off the MCB or remove the fuse from the consumer unit. Unplug all appliances from the ring and switch off any fixed appliances connected to it.

Remove the socket, loosen the live terminal and separate the two red conductors. Leave the other wires in place. With one probe of the continuity tester touching the socket's neutral terminal, place the other probe on the bared end of each red wire in turn. The tester's indicator should not light up in either case – provided you have unplugged everything and switched off fixed appliances, as described above.

Now touch one probe against the end of one of the red conductors, and the other probe against the end of the other red conductor. If the indicator lights up, you can be sure it is a ring circuit and you can safely add your spur.

Connecting to an existing socket

Fix the new socket, then wire it up in the normal way (see opposite) and run its spur cable to the existing socket outlet. Switch off the electricity and remove the existing socket. You may have to enlarge the entry hole, or knock out another one, to take the spur cable. Feed the cable into the box, prepare the conductors, and insert their bared ends together with those of the conductors of the ring circuit. Insert the wires in their terminals (brown/red – L; blue/black – N; and green-and-yellow – E) and replace the socket. Then switch the power on and test the new socket.

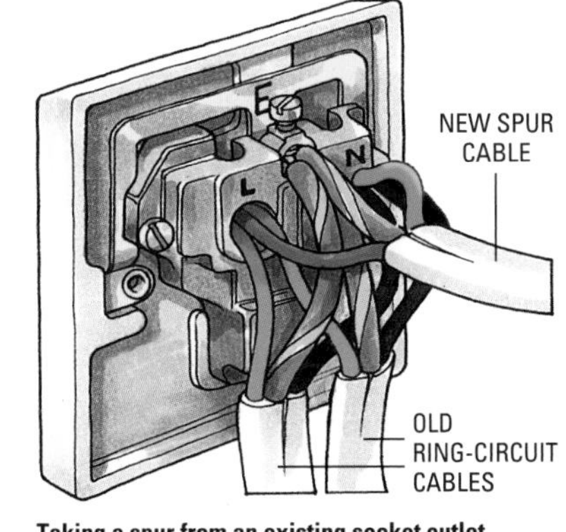

Taking a spur from an existing socket outlet

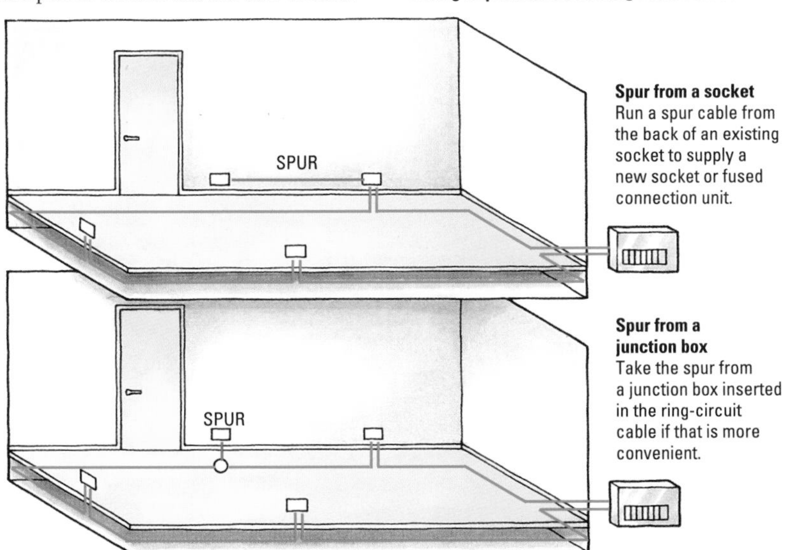

Spur from a socket
Run a spur cable from the back of an existing socket to supply a new socket or fused connection unit.

Spur from a junction box
Take the spur from a junction box inserted in the ring-circuit cable if that is more convenient.

CONNECTING TO A JUNCTION BOX

You will need a 30amp junction box with three terminals to connect to a ring circuit. It will have either knock-out cable-entry holes or a special cover that rotates to blank off unneeded holes. The cover must be screw-fixed. Lift a floorboard close to the new socket, so you can connect to the ring-circuit cable without stretching it.

Making a platform

Fix a platform for the box by nailing battens near the bottom of two joists (see right) and screwing a 100 x 25mm (4 x 1in) strip of wood between the joists and resting on the battens. Loop the ring-circuit cable over the platform before fixing it, so that the cable need not be cut for connecting up. Remove the cover, screw the junction box to the platform, and break out two cable-entry holes. If you do forget to loop the cable over the platform, simply cut the cable when you come to connect it up.

Connecting the ring-circuit cable

Turn off the power at the consumer unit, then rest the ring-circuit cable across the box and mark the amount of sheathing to remove. Slit it lengthwise and peel it off the conductors. Don't cut the live and neutral conductors, but slice away just enough insulation on each to expose a section of bare wire that will fit into a terminal (see right). Cut the earth wire and fit insulating sleeves on the two ends.

Remove the screws from all three of the terminals and lay the wires across them – with the earth wire in the middle terminal, and the live and neutral ones on each side. Push the wires home with a screwdriver.

Connecting the spur

Having fitted and wired the new spur socket, run its cable to the junction box. Cut and prepare the ends of the wires, and break out an entry hole so the spur wires can be fitted to the terminals of the box (see right). Attach the new brown wire to the terminal holding the old red ones, and the new blue wire to the terminal holding the old black ones. Connect all earth wires to the central terminal. Replace the fixing screws – starting them by hand as they are easily cross-threaded, then tightening them up with a screwdriver. Check that all the wires are secured and that the cables all fit snugly in their entry holes, with the sheathing running into the box; then fit the cover on the box.

Fix each cable to a nearby joist with cable clips, to take the strain off the terminals, then replace the floorboards.

Switch the power back on and test the new socket.

• Old colour coding
When working on a house built before 2005 you are likely to find that the existing cables are colour-coded black for neutral and red for live. The diagrams on this page show new-style spur cables being connected to old-style circuit cables.

Make a wooden platform for a junction box

OLD RING-CIRCUIT CABLE
NEW SPUR CABLE
OLD RING-CIRCUIT CABLE

Taking a spur from a junction box

Extending a ring circuit

There are times when it's better to extend a ring circuit than to fit spurs – for example, if you want to wire a room that isn't adequately serviced, or all of the conveniently placed sockets already have spurs running from them. You can break into the ring at an existing socket or via junction boxes.

● Switching off
However you plan to extend a ring circuit, remember to switch off the power at the consumer unit before you break into the ring.

Using an existing socket

Disconnect one of the in-going cables from a socket on the ring circuit and take it to the first new socket. Do this via a junction box if the cable won't otherwise reach. Continue the extension with a new section of cable, running it from socket to socket – finally running it from the last new socket back to the one where you broke into the ring. Joining the new cable to the old cable within the socket completes the circuit.

Using junction boxes

Cut the ring cable and connect each cut end to a junction box, then run a new length of cable from one box to the other, looping it into the new sockets.

Running the extension

No matter how you plan to break into the ring, always install the new cable first and then connect it up to the circuit at the last moment. This allows you to use power tools to run the extension – but don't forget to switch the power off just before connecting up.

Decide positions for the new sockets and plan your cable run (an easy route is preferable to a shorter but more difficult one), allowing some slack in the cable.

Cut out the plaster and brickwork for sockets and cable, then fit the boxes for the sockets. Now run the cable, leaving enough spare for joining to the ring circuit, and take it up behind the skirting to the first socket. Leave a loop hanging near the box (see right), then take the cable on to the next one – and so on till all the new sockets are supplied. Take the excess cable on to the point where you plan to join the ring.

Fit the new sockets; then switch off the electricity, break into the ring, and connect the extension to it. Switch the power on and test all the new sockets separately. Make good the plasterwork.

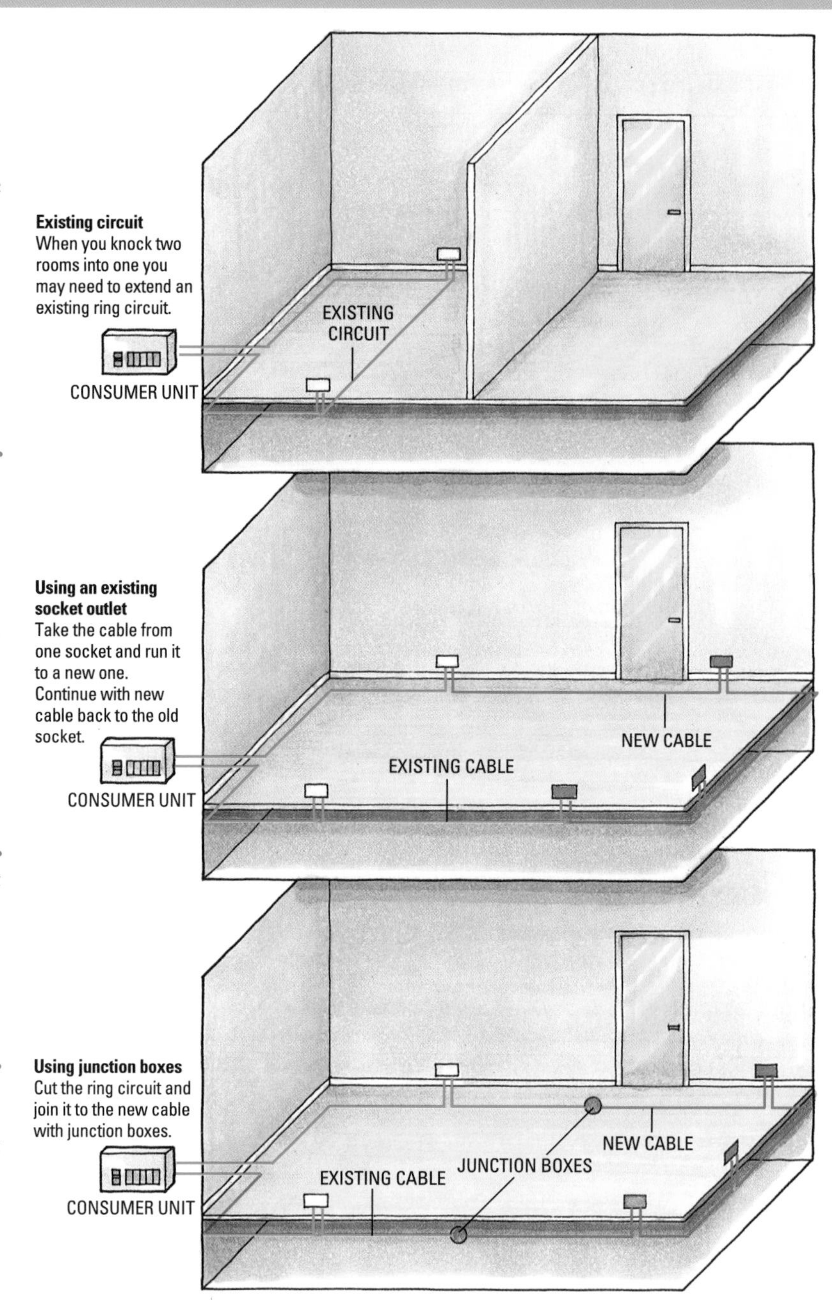

Existing circuit
When you knock two rooms into one you may need to extend an existing ring circuit.

Using an existing socket outlet
Take the cable from one socket and run it to a new one. Continue with new cable back to the old socket.

Using junction boxes
Cut the ring circuit and join it to the new cable with junction boxes.

LEAVE SOME SLACK IN THE CIRCUIT

Don't pull the cable too tight when you're running a new circuit. It places a strain on the connections and will make it difficult to modify the circuit at a later stage, should that become necessary.

Leave a generous loop of cable at each of the new socket positions until you have run the complete circuit. At that stage you can pull the loop back, ready for connecting to the socket.

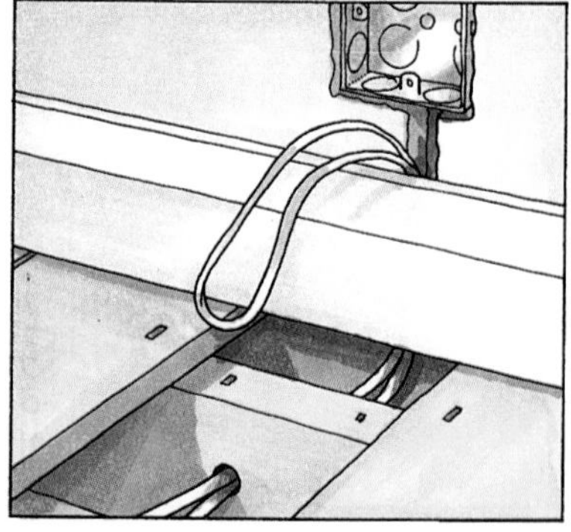

Leave ample cable above the skirting

☛ SEE ALSO: Switching off 16, Ring-circuit regulations 21, Running cable 23–5, Positioning sockets 28, Mounting boxes 28–9, Wiring sockets 30, Junction box 31, Circuit lengths 66

Converting a radial circuit Ⓝ

If you have a radial circuit, you may want to convert it to a ring circuit, particularly if you need to supply a larger area. Before starting work, switch off at the consumer unit.

Checking cable and fuse

If the radial circuit is wired with $2.5mm^2$ cable (solid conductors), continue the circuit back to the consumer unit with the same size cable but substitute a 30amp fuse and fuseway in place of the 20amp fuse. If the original radial circuit is wired in $4mm^2$ cable, continue using the same size cable for the remainder of the ring. Check there's a 30amp fuse.

The extra cable is run in exactly the same way as described for extending a ring circuit (see opposite). Join the new cable at the last socket on the radial circuit and run it to all the new sockets. From the last socket, run the cable to the consumer unit.

At the consumer unit

You should examine your consumer unit and familiarize yourself with it. But even when the unit is switched off, the cable connecting the meter to the main switch is still live – so take great care.

First locate the terminals to which the radial circuit is connected. The live (red wire) terminal is on the fuseway (or MCB) from which you removed the circuit fuse prior to starting work. The neutral (black wire) terminal is on the neutral block, to which all of the black wires are connected. You can usually trace the black wire you are looking for by working along from the sheathed part of the cable. Similarly, you can locate the earth terminal by tracing the green-and-yellow-insulated conductor.

Pass the new cable into the consumer unit close to the original radial-circuit cable. Cut the new cable to length, then strip off the sheathing and prepare the conductors.

Disconnect the live (red) conductor from its terminal and, having checked for continuity (see far right), put it back into the terminal along with the brown wire from the new cable. Do the same for the black and blue wires and then the green-and-yellow ones, slipping a sleeve over the new earth wire.

Check that the circuit fuse is of the correct rating, then replace the fuse carrier. Close the consumer unit, switch on the power, and test the circuit.

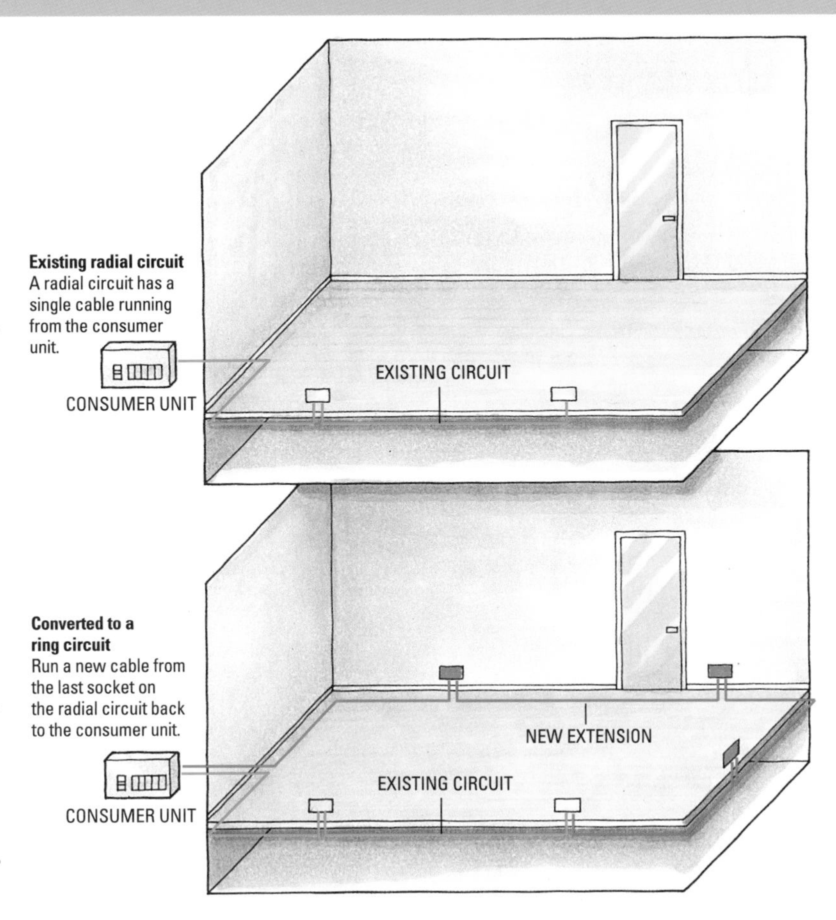

Existing radial circuit
A radial circuit has a single cable running from the consumer unit.

Converted to a ring circuit
Run a new cable from the last socket on the radial circuit back to the consumer unit.

● Testing for continuity
Check the continuity of the new ring circuit before you insert the conductors into their terminals in the consumer unit. Using a continuity tester, place one of its probes on the live conductor at one end of the circuit cable, and its other probe on the live conductor at the other end of the ring. Press the tool's test button, and if the circuit is complete the tester's indicator will illuminate. Carry out the same test for the neutral conductors, then for the earth wires.

CONNECTING TO THE CONSUMER UNIT

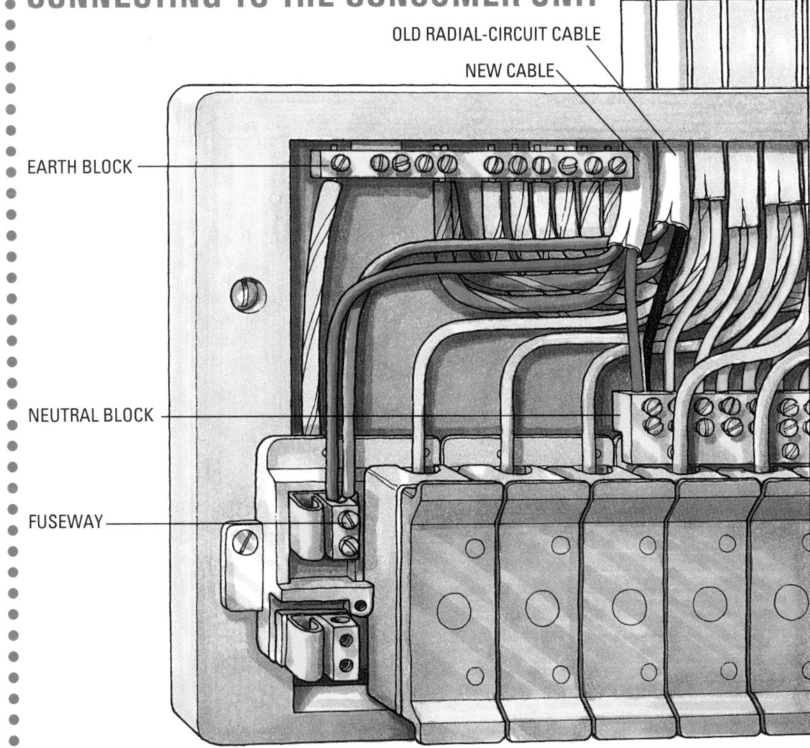

Wire the new cable and the old radial-circuit cable into the same terminals

● Old colour coding
When working on a house built before 2005 you are likely to find that the existing cables are colour-coded black for neutral and red for live. The diagram (left) shows new-style cable being connected to existing old-style circuit cable.

SEE ALSO: Switching off 16, Consumer units 18, Fuse ratings 19, Radial-circuit regulations 21, Cable 22, Running cable 23–5, Positioning sockets 28, Wiring sockets 30, Circuit lengths 66, Final tests 67

Fixed appliances

Socket outlets are designed to enable appliances to be moved from room to room. As a result, a socket may be used for various appliances at different times. But many electrical appliances are fixed to the structure of the house, or stand in one position all the time. Such appliances may therefore just as well be wired into your electrical installation permanently. Indeed in some cases there is no alternative, and some require radial circuits of their own direct from the consumer unit.

FUSED CONNECTION UNITS

A fused connection unit is basically a device for joining the flex (or sometimes cable) of an appliance to circuit wiring. The connection unit incorporates the added protection of a cartridge fuse similar to that found in a 13amp plug. If the appliance is connected by a flex, choose a unit that has a cord outlet in the faceplate.

Some fused connection units are fitted with a switch, and some of these have a neon indicator that shows at a glance whether they are switched on. A switched connection unit allows you to isolate the appliance from the mains.

All fused connection units are single (there are no double versions available) with square faceplates that fit metal boxes for flush mounting or standard surface-mounted plastic boxes.

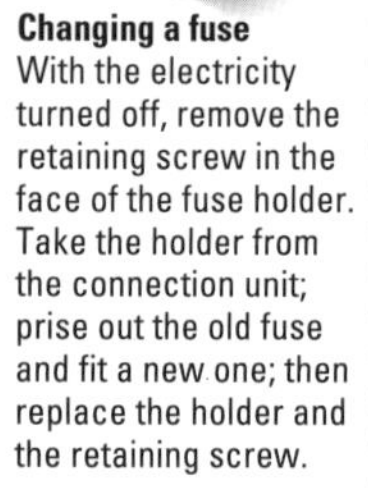

Changing a fuse
With the electricity turned off, remove the retaining screw in the face of the fuse holder. Take the holder from the connection unit; prise out the old fuse and fit a new one; then replace the holder and the retaining screw.

Fused connection units
1 Unswitched connection unit.
2 Switched unit with cord outlet and indicator.
3 Connection unit and socket outlet in a dual mounting box.

Small appliances

Small permanent electrical appliances with ratings of up to 3000W (3kW) – wall heaters, heated towel rails, cooker hoods and so on – can be wired into a ring or radial circuit by means of fused connection units.

Although such appliances could be connected by means of 13amp plugs to socket outlets, the electrical contact would not be so good – and there is also some risk of fire with that type of permanent installation.

Before wiring a fused connection unit to the house circuitry, always remember to switch off the power at the consumer unit.

Mounting a fused connection unit

A fused connection unit is mounted in the same type of box as an ordinary socket outlet, and the box is fixed to the wall in exactly the same way. The unit can also be mounted in a dual box that is designed to hold two single units – for example, a standard socket outlet beside a connection unit. The socket is wired to the ring circuit, and the two units are linked together inside the box by a short 2.5mm² spur.

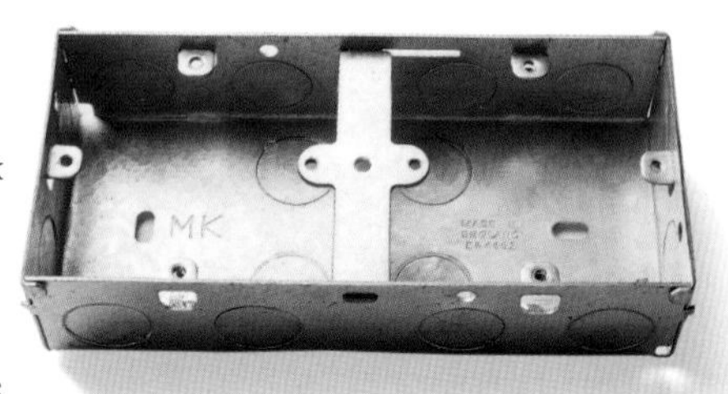

Dual mounting box

Wiring a fused connection unit

Fused connection units can be supplied by a ring circuit, a radial circuit or a spur. Some appliances are connected to the unit with flex, others with cable. Either way, the wiring arrangement inside the unit is the same. Units with cord outlets have clamps to secure the connecting flex.

An unswitched connection unit has two live (L) terminals – one marked 'Load' for the brown wire of the flex, and the other marked 'Mains' for the brown or red wire from the circuit cable. The blue wire from the flex and the blue or black wire from the circuit cable go to similar neutral (N) terminals; and both earth wires are connected to the unit's earth (E) terminal or terminals **(1)**.

Switched connection unit

A fused connection unit with a switch also has two sets of terminals. Those marked 'Mains' are for the spur or ring cable that supplies the power; the terminals marked 'Load' are for the flex or cable from the appliance.

Wire up the flex side first, connecting the brown wire to the L terminal, and the blue one to the N terminal, both on the 'Load' side. Connect the green-and-yellow wire to the E terminal **(2)** and tighten the cord clamp.

Attach the circuit conductors to the 'Mains' terminals – brown or red to L, and blue or black to N; then sleeve the earth wire and take it to the E terminal **(2)**.

If the fused connection unit is on a ring circuit, you must fit two circuit conductors into each 'Mains' terminal and into the earth terminal. Before securing the unit in its box with the fixing screws, make sure the wires are held firmly in the terminals and can fold away neatly.

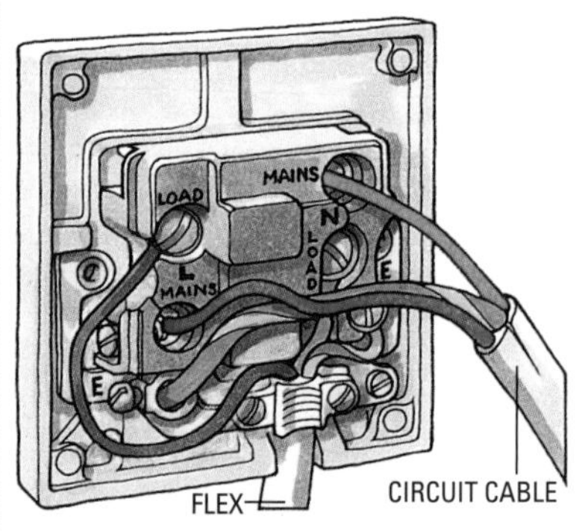

1 Wiring a fused connection unit

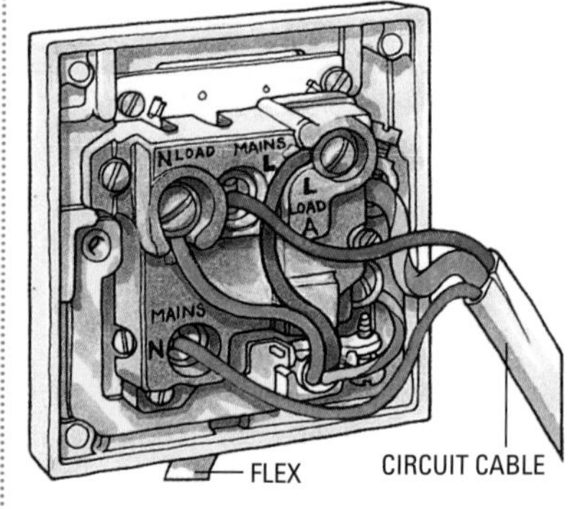

2 Wiring a switched fused connection unit

SEE ALSO: Stripping flex 13, Switching off 16, Power circuits 21, Stripping cables 22, Mounting boxes 28–9

Wiring heaters

When you're installing a skirting heater or wall-mounted heater or an oil-filled radiator, wire the appliance to a fused connection unit mounted nearby, at a height of about 150 to 300mm (6in to 1ft) from the floor. Whether the connection to the unit is by flex or cable will depend on the type of appliance. Follow the manufacturer's instructions for wiring, and fit the appropriate fuse in the connection unit.

In a bathroom, a fused connection unit must be mounted outside zones 0 to 3. Any heater that is mounted near the floor of a bathroom must therefore be wired to a connection unit installed outside the room. If the appliance is fitted with flex, mount a flexible-cord outlet **(1)** next to the appliance – and then run a cable from the outlet to the fused connection unit outside the bathroom and connect it to the 'Load' terminals in the unit.

The flexible-cord outlet is mounted either on a standard surface-mounted box or flush on a metal box. At the back of the faceplate are three pairs of terminals to take the conductors from the flex and the cable **(2)**.

Radiant wall heaters Ⓝ

Radiant wall heaters for use in bathrooms must be fixed high on the wall, outside zones 0 to 2. A fused connection unit fitted with a 13amp fuse (or 5amp fuse for a heater of 1kW or less) must be mounted at a high level outside the zones, and the heater must be controlled by a double-pole pull-cord switch (with this type of switch, both live and neutral contacts are broken when it is off). Many heaters have a built-in double-pole switch; otherwise, you must fit a ceiling-mounted 15amp double-pole switch between the fused connection unit and the heater. Switch terminals marked 'Mains' are for the cable on the circuit side of the switch; those marked 'Load' are for the heater side. The earth wires are connected to a common terminal on the switch box.

If it is not possible to run a spur to the fused connection unit from a socket outside the bathroom, run a separate radial circuit from the connection unit to a 15amp fuseway in the consumer unit, using 2.5mm^2 cable. In either case, the circuit should be protected by a 30 milliamp RCD.

Heated towel rail Ⓝ

The Wiring Regulations covering other kinds of heater also apply to a heated towel rail situated in a bathroom. As the towel rail is mounted near the floor, run a flex from it to a flexible-cord outlet, which must in turn be wired to a fused connection unit outside the bathroom. For a towel rail of 1kW or less fit a 5amp fuse; otherwise fit a 13amp fuse.

If a heated towel rail is installed in a bedroom, the fused connection unit can be mounted alongside it.

Heat/light unit Ⓝ

Heat/light units, which are sometimes fitted in bathrooms, incorporate a radiant heater and a light fitting in the one appliance. Although they are ceiling-mounted, usually in the position of the ceiling rose, these units must never be connected to lighting circuits.

To install a heat/light unit in this position, turn off the power and, having identified the lighting cables, remove the rose and withdraw the cables into the ceiling void. Fit a junction box to a nearby joist and terminate the lighting cables at that point **(3)**. Don't connect the switch cable, as it won't be needed.

Run a 2.5mm^2 two-core-and-earth spur cable from an unswitched fused connection unit mounted outside the bathroom to a ceiling-mounted 15amp double-pole switch, and from there to the heat/light unit.

Connect up to the fused connection unit (see opposite), and then wire the heat/light unit according to the maker's instructions. Fit a 13amp fuse in the connection unit.

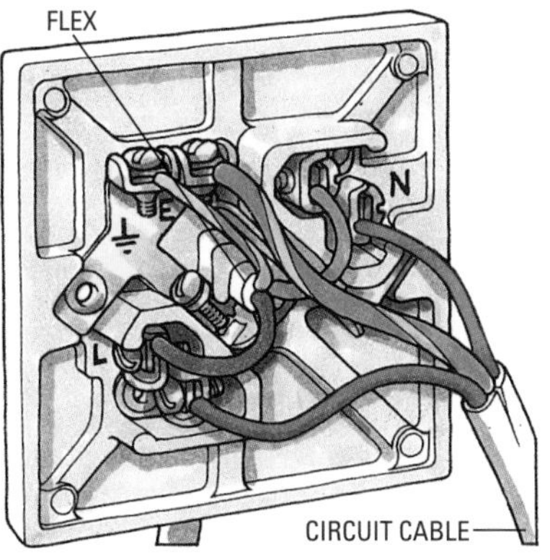

2 Wiring a flexible-cord outlet

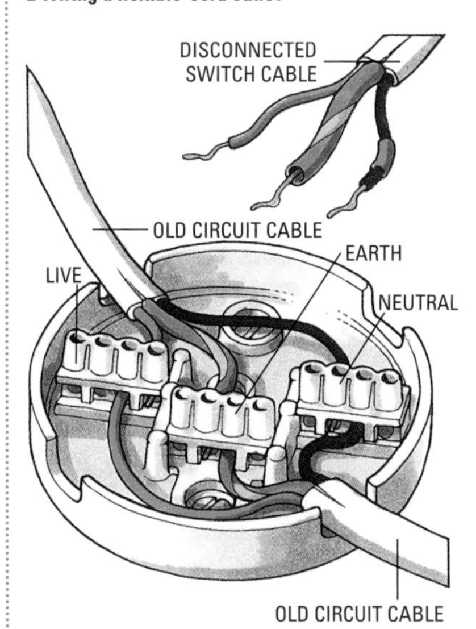

3 Terminating the lighting cables
Join the circuit cables in a junction box. Label the disconnected switch wire for future reference.

● Old colour coding
When working on a house built before 2005 you are likely to find that the existing cables are colour-coded black for neutral and red for live. The diagrams on this page show new-style cables being connected to existing old-style circuit cables.

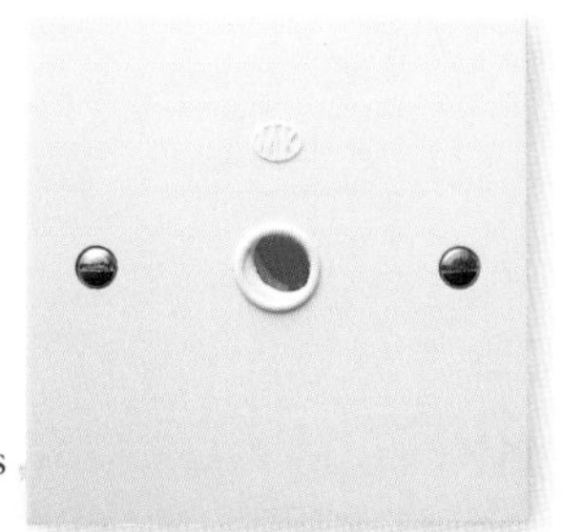

1 Flexible-cord outlet

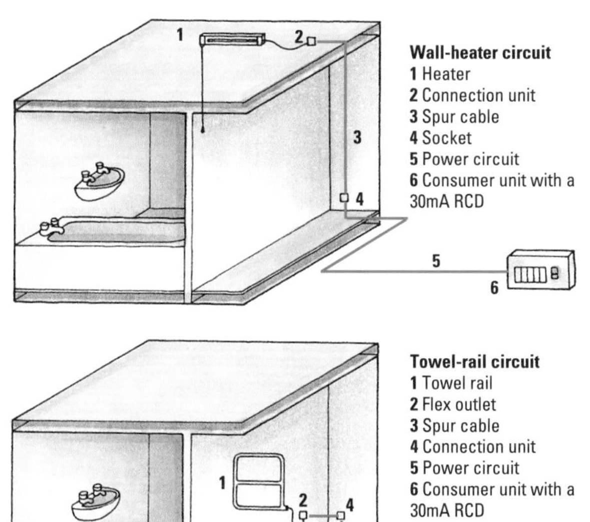

Wall-heater circuit
1 Heater
2 Connection unit
3 Spur cable
4 Socket
5 Power circuit
6 Consumer unit with a 30mA RCD

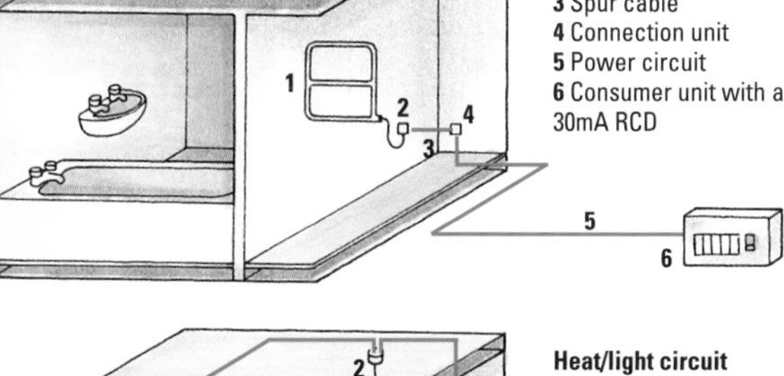

Towel-rail circuit
1 Towel rail
2 Flex outlet
3 Spur cable
4 Connection unit
5 Power circuit
6 Consumer unit with a 30mA RCD

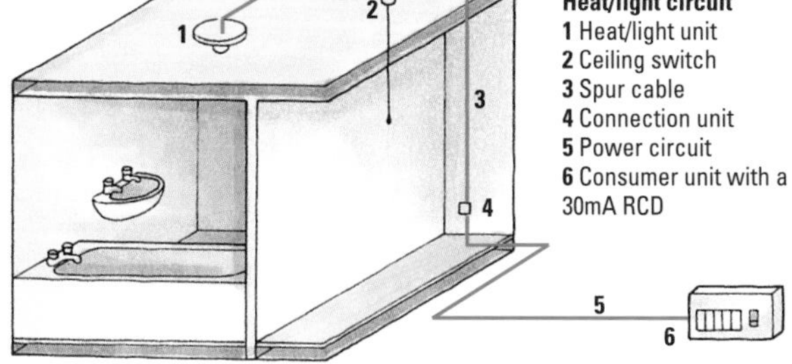

Heat/light circuit
1 Heat/light unit
2 Ceiling switch
3 Spur cable
4 Connection unit
5 Power circuit
6 Consumer unit with a 30mA RCD

● RCD protection
When installing any electrical appliance in a bathroom, the circuit should be protected by a 30 milliamp RCD.

☛ SEE ALSO: **Building Regulations 8, Fuses 9, 19, Bathroom safety 10, Zones for bathrooms 11, Switching off 16, Running cable 23–5, Double-pole ceiling switch 47, Ceiling-rose connections 48**

Wiring small appliances

Extractor fan

To install an extractor fan in a kitchen, mount a fused connection unit 150mm (6in) above the worktop and run a cable to the fan or to a flexible-cord outlet next to it. If the fan has no integral switch, use a switched connection unit to control it. Fit a 3 or 5amp fuse, as recommended by the manufacturer.

If the fan's speed and direction are controllable, it may have a separate control unit – in which case you need to wire the connection unit to the control unit, following the maker's instructions.

To install an extractor fan in a bathroom, mount the fused connection unit outside the room and run the cable to the fan or flex outlet via a double-pole switch mounted on the ceiling.

• Fan in a bathroom
In a bathroom, an extractor fan must be mounted outside zones 0 and 1.

Fridges, dishwashers and washing machines

There is no reason why you cannot plug an appliance like a fridge, dishwasher or washing machine into a standard socket outlet – except that in a modern kitchen such appliances are installed under worktops, and sockets mounted behind them are difficult to reach. It's therefore generally more convenient to mount a switched fused connection unit 150mm (6in) above the worktop, then connect it to the ring circuit and run a spur – using $2.5mm^2$ cable – from the connection unit to a socket outlet mounted behind the appliance.

• RCD protection
When installing any electrical appliance in a bathroom, the circuit should be protected by a 30 milliamp RCD.

Cooker hood

Either mount a fused connection unit, fitted with a 3amp fuse, close to the cooker hood or mount the connection unit at worktop height and then run a $1mm^2$ cable from the unit to a flexible-cord outlet beside the hood.

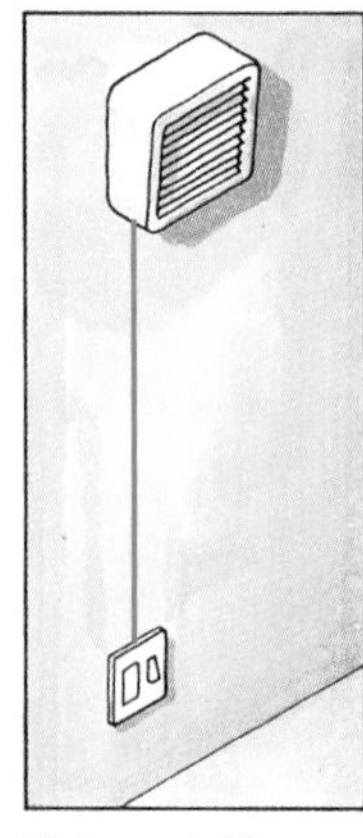

Wall-mounted fan
Run a $1.5mm^2$ cable from a fused connection unit to a wall-mounted extractor fan.

Instantaneous water heater

To provide an on-the-spot supply of hot water, you can install an instantaneous water heater above a washbasin or sink. Join a 3kW model by heat-resistant flex to a switched fused connection unit mounted out of reach of water splashes from the basin or sink.

If the heater is for use in a bathroom, wire it via a flex outlet to a ceiling pull-switch and then to a connection unit outside the bathroom. The connection unit must be fitted with a 13amp fuse.

Wire a 7kW water heater in the same way as a shower. If it is situated in the kitchen, you can use a double-pole wall switch to control it.

Waste-disposal unit

A waste-disposal unit is housed in the cupboard unit below the sink. Mount a switched fused connection unit 150mm (6in) above a worktop near the sink, but well out of reach of small children and splashes from the sink. From the unit, run a $1mm^2$ cable to a flex outlet next to the waste-disposal unit. Clearly label the connection unit 'WASTE DISPOSAL', to avoid accidents. Fit a 13amp fuse.

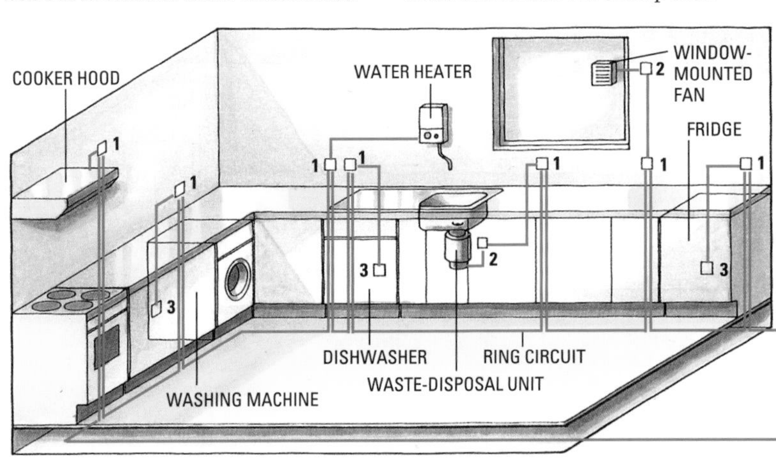

Circuits for kitchen equipment
1 Connection units
2 Flex outlets
3 Socket outlets

SHAVER SOCKETS

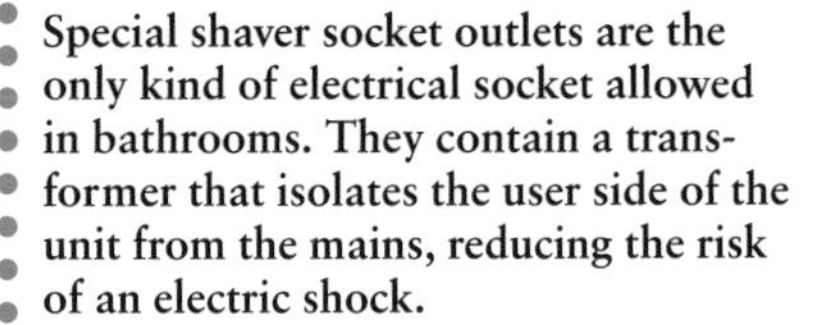
Special shaver socket outlets are the only kind of electrical socket allowed in bathrooms. They contain a transformer that isolates the user side of the unit from the mains, reducing the risk of an electric shock.

This type of socket has to conform to the exacting British Standard BS EN 60742 Chapter 2, Section 1. However, there are shaver sockets that do not have an isolating transformer and therefore don't conform to this standard. These are quite safe to install and use in a bedroom – but this type of socket must not be fitted in a bathroom.

You can wire a shaver socket from a junction box on an earthed lighting circuit or from a fused connection unit, fitted with a 3amp fuse, on a ring-circuit spur. If you're installing the shaver socket in a bathroom, then the fused connection unit must be positioned outside the room. Run $1mm^2$ two-core-and-earth cable from the connection unit to the shaver socket; then connect the conductors: brown to L and blue to N **(1)**. Sheath the earth wire with a green-and-yellow sleeve and connect it to E.

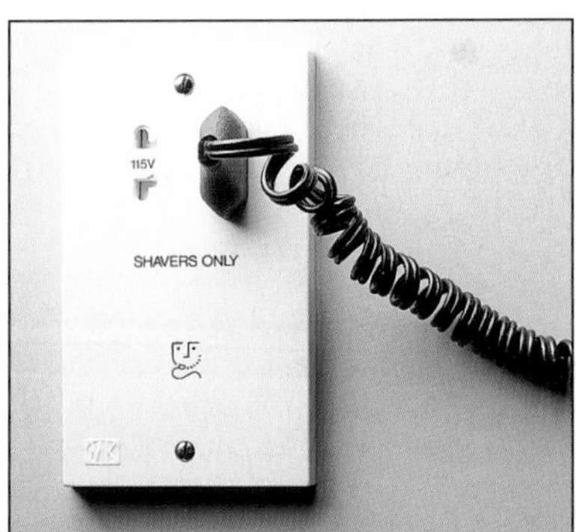

Shaver unit for use in a bathroom

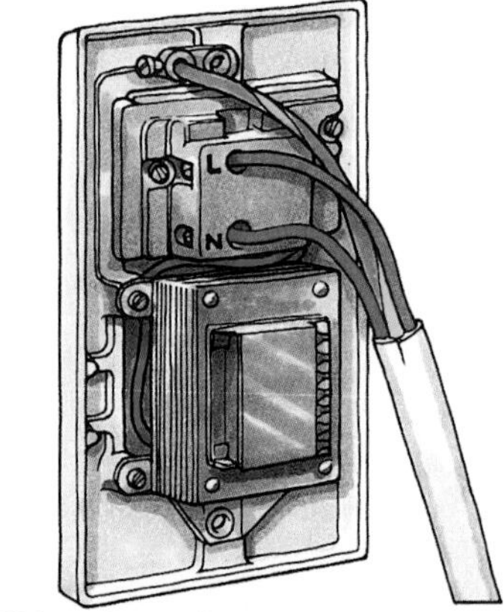

1 Wiring a shaver unit

SEE ALSO: Building Regulations 8, Zones 11, Running a spur 31, Fused connection units 34, Flex outlet 35, Double-pole switch 39, 47, Connecting to a light circuit 47, Circuit lengths 66

Wiring a cooker (N)

Appliances, such as cookers, that have a power load greater than 3000W (3kW) must have their own radial circuits connected directly to the consumer unit, with separate fuses protecting them.

Cooker circuits

Small table cookers and separate ovens that rate no more than 3kW can be connected to a ring circuit by a fused connection unit or even by means of a 13amp plug and socket. However, most cookers are much more powerful and must be installed on their own circuits.

The radial circuit

A cooker must be connected to a radial circuit – a single cable that runs back to the consumer unit. Between the cooker and the consumer unit, you must install a cooker control unit (which is basically a double-pole isolating switch). Some cooker control units incorporate a single 13amp socket outlet that can be used for appliances such as an electric kettle.

When cookers up to 13.5kW are connected to a control unit that has a socket, the radial circuit must be run using $4mm^2$ two-core-and-earth cable and it must be protected by a 30amp fuse or a 32amp MCB. Larger cookers, up to 18kW, can be connected to a similar circuit, but you must use $6mm^2$ two-core-and-earth cable and a 40amp MCB – you cannot use a fuse. (See CIRCUITS: MAXIMUM LENGTHS).

With either of the circuits described above, it's safe to use a unit that does not have a socket outlet. In fact, if you use a socketless unit, the Wiring Regulations allow you to run longer circuit lengths and to use a fuse with the larger cookers (instead of an MCB). If either of these is desirable, consult an electrician.

Consumer unit or switchfuse unit

You can either make use of a spare fuse-way in your existing consumer unit or fit a separate switchfuse unit or a single-way consumer unit with an MCB (this performs the same sort of function as an ordinary consumer unit but for a single appliance). If you use a switchfuse unit, make sure it can take a cartridge fuse.

Positioning the cooker control unit

The control unit must be situated within 2m (6ft 6in) of the cooker. The unit has to be easily accessible – so don't install it inside a cupboard or under a worktop.

A single control unit can serve both sections of a split-level cooker, with separate cables running to the hob and the oven, provided that the control unit is within 2m (6ft 6in) of both parts. (If this isn't possible with your cooker, you will need to install a separate control unit for each part.) The connecting cables must be of the same size as the cable used in the radial circuit.

A freestanding cooker will have to be moved from time to time for cleaning, so wire it with sufficient cable to allow it to be moved well out from the wall. The cable is connected to a terminal outlet box, which is screwed to the wall about 600mm (2ft) above floor level. A fixed cable runs from the outlet box to the cooker control unit.

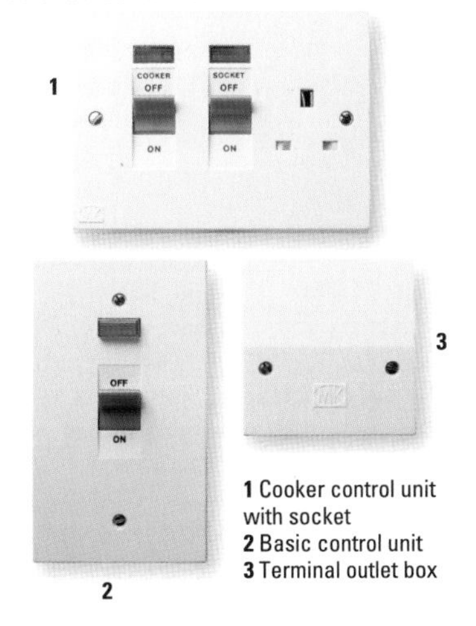

1 Cooker control unit with socket
2 Basic control unit
3 Terminal outlet box

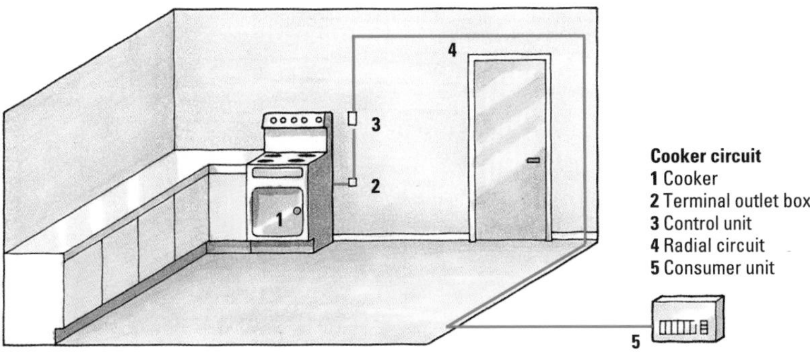

Cooker circuit
1 Cooker
2 Terminal outlet box
3 Control unit
4 Radial circuit
5 Consumer unit

WIRING THE CONTROL UNIT

Having decided on the position for your control unit, if it's going to be surface-mounted simply knock out the cable-entry holes in the mounting box and screw it to the wall. If it's to be flush-mounted, cut a hole in the plaster and brickwork for the metal box.

Running cable

Run and fix the cable, taking the most economical route to the cooker from the consumer unit or switchfuse unit.

If you're going to bury the cable in the plaster, cut a chase in the wall up to the cooker control unit, then cut similar chases for cables running to the separate hob and oven of a split-level cooker or for a single cable running to a terminal outlet box.

Connecting up the control unit

Feed the circuit cable and cooker cable into the control unit, then strip and prepare the conductors for connection.

There are two sets of terminals in the control unit: one marked 'Mains' for the circuit conductors, and the other marked 'Load' for the cooker cable. Run the brown wires to the L terminals, and the blue ones to the terminals marked N. Put green-and-yellow sleeves on both earth conductors and connect them to the E terminal **(1)**. Screw the faceplate to the mounting box.

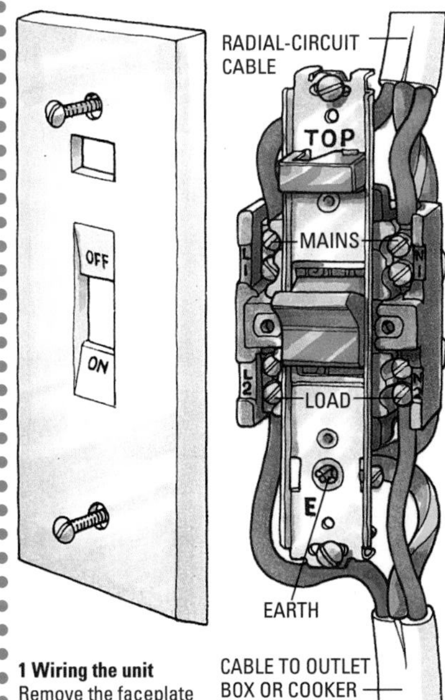
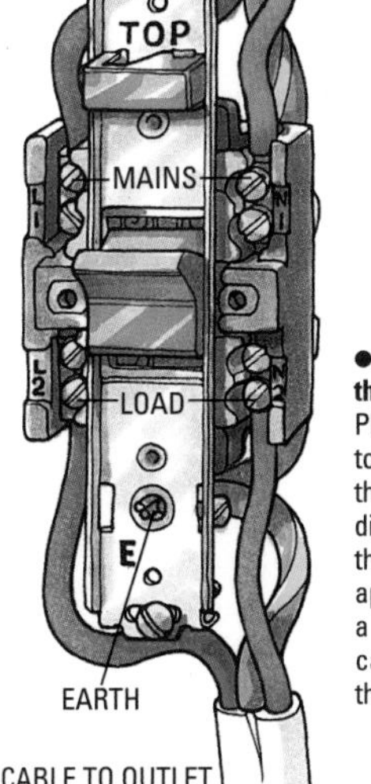

1 Wiring the unit
Remove the faceplate to wire some units.

• A safe position for the cooker control unit
Place the control unit to the right or left of the cooker – but never directly above it. Ensure that the flex from an appliance plugged into a control unit's socket cannot drape across the cooker.

Wiring to the cooker

When you connect the cable to the oven and the hob, follow the manufacturer's instructions exactly.

For a freestanding cooker, run the cable down the wall from the cooker control unit to the terminal outlet box, which has terminals for connecting both of the cables. Strip the wires of the control-unit cable and insert them in the terminals **(1)**, then insert the wires of the cooker cable in the same terminals, matching colour for colour, and secure it with the clamp. Screw the plastic faceplate onto the outlet box.

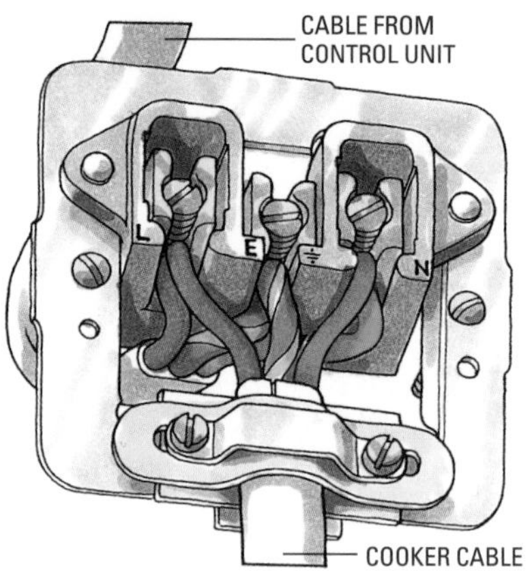

1 Wiring a terminal outlet box

Wiring a switchfuse unit

If you are wiring to a fuseway in your consumer unit, run the brown wire to the terminal on the fuseway, the blue one to the neutral block, and – having first sleeved it – the earth wire to the earth block. All the other connections will already have been made. Don't forget to switch off the power before starting this work, and remember that even then the cable connecting the meter to the main switch is still live.

Here we will assume that the cooker circuit is to be run from a switchfuse unit. Screw the unit to the wall, close to the consumer unit. Feed the cooker-circuit cable into it, and prepare the conductors for connection. Fix the brown wire to the live terminal on the fuseway (MCB in a single-way consumer unit); the blue wire to the neutral terminal; and the sleeved earth wire to the earth terminal **(2)**.

Prepare the meter leads, one blue and one brown, from PVC-sheathed-and-insulated $16mm^2$ single-core cable. (Use $10mm^2$ cable if $16mm^2$ cable is too thick for the switchfuse-unit terminals, but keep the meter leads as short as possible.) Bare about 25mm (1in) of each cable and connect the leads to their separate terminals on the main isolating switch – brown to L, and blue to N **(2)**. For an earth lead, prepare a similar length of the same-size single-core cable sheathed in green-and-yellow PVC and attach it to the earth terminal in the switchfuse unit **(2)** in readiness for connection to the consumer's earth terminal. Don't make the connection to the electricity company's earth yourself.

Fit the appropriate fuse, then plug in the fuse carrier. Finally, label the carrier to indicate which circuit is run from the unit and fit the cover.

2 Wiring a switchfuse unit for the cooker

Connecting to the mains

A new circuit has to be tested by a competent electrician, whose certificate stating that the wiring complies with the Wiring Regulations must be submitted to the electricity company when you apply for connection to the mains. Don't attempt to make this connection (which has to be made via the meter) yourself.

It may not be possible to attach both sets of meter leads – from the consumer unit and the new switchfuse unit – to the meter, and you may have to install a connector block that has enough terminals to accommodate all the conductors. The electricity company will do this for a fee (before starting it's advisable to consult the company about these matters).

Hot water

The water in a storage cylinder can be heated by an electric immersion heater, providing a central supply of hot water for the whole house.

The heating element is rather like a larger version of the one that heats an electric kettle. It is normally sheathed in copper, but more expensive sheathings of incoloy or titanium will increase the life of the element in hard-water areas.

Adjusting the water temperature
The thermostat that controls the maximum temperature of the water is set by adjusting a screw inside the plastic cap covering the terminal box **(1)**.

Types of immersion heater
An immersion heater can be installed either from the top of the cylinder or from the side, and top-entry units can have single or double elements.

With the single-element top-entry type, the element extends down almost to the bottom of the cylinder, so that all of the water is heated whenever the heater is switched on **(2)**.

For economy, one of the elements in the double-element type is a short one for daytime top-up heating, while the other is a full-length element that heats the entire contents of the cylinder, using the cheaper night-rate electricity **(3)**. A double-element heater that has a single thermostat is called a twin-element heater; one with a thermostat for each element is known as a dual-element heater.

Side-entry elements are of identical length. One is positioned near to the bottom of the cylinder, and the other a little above halfway up **(4)**.

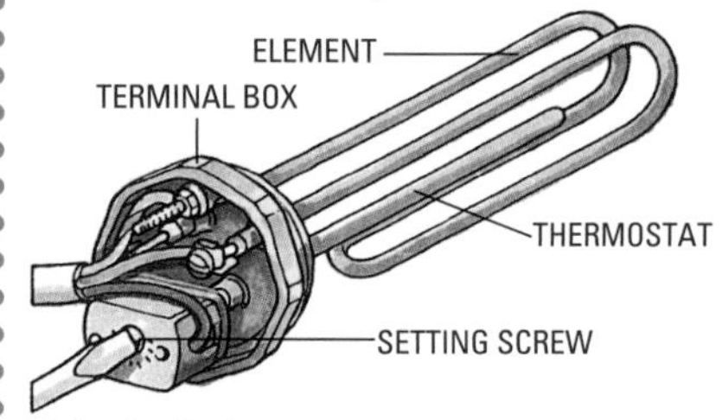

1 Adjusting the thermostat

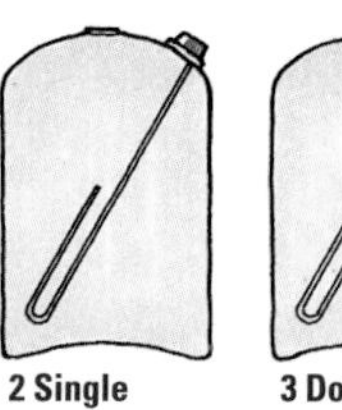

2 Single element

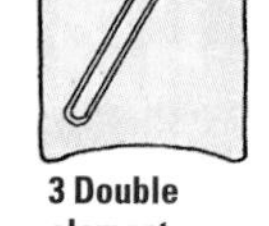

3 Double element

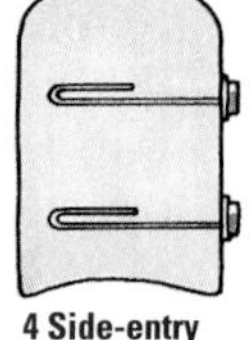

4 Side-entry elements

SEE ALSO: Building Regulations 8, Switching off 16, Consumer units 18, Circuit fuses 19, Cables 22, Stripping cable 22, Running cable 23–5, Final tests 67

Wiring immersion heaters

If you agree to their installing a special meter, your electricity company will supply you with cheap-rate power for seven hours sometime between midnight and 8.00 a.m., the exact period being at the discretion of the company. This scheme is called Economy 7.

Provided you have a cylinder that is large enough to store hot water for a day's requirements, you can benefit by heating all your water during the Economy 7 hours. Even if you heat your water electrically only in summer, the scheme may be worthwhile. For the water to retain its heat all day, you must have an efficient insulating jacket fitted to the cylinder or a cylinder already factory-insulated with a layer of heat-retaining foam.

If your cylinder is already fitted with an immersion heater, you can use the existing wiring by fitting an Economy 7 programmer, a device that will switch your immersion heater on automatically at night and heat up the whole cylinder. Then if you occasionally run out of hot water during the day, you can always adjust the programmer's controls to boost the temperature briefly, using the more expensive daytime rate.

You can make even greater savings if you have two side-entry immersion heaters or a dual-element one. The programmer will switch on the longer element, or the bottom one, at night; but if the water needs heating during the day, then the upper or shorter element is used.

Economy 7 without a programmer

You can have a similar arrangement without a programmer if you wire two separate circuits for the elements. The upper element is wired to the daytime supply, while the lower one is wired to its own switchfuse unit and operated by the Economy 7 time switch during the hours of the night-time tariff only. A setting of 75°C (167°F) is recommended for the lower element, and 60°C (140°F) for the upper one. If your water is soft or your heater elements are sheathed in titanium or incoloy, you can raise the temperatures to 80°C (175°F) and 65°C (150°F) respectively without reducing the life of the elements.

To ensure that you never run short of hot water, leave the upper unit switched on permanently. It will only start heating up if the thermostat detects a temperature of 60°C (140°F) or less, which should happen very rarely if you have a large cylinder that is properly insulated.

The circuit

The majority of immersion heaters are rated at 3kW; but although you can wire most 3kW appliances to a ring circuit, an immersion heater is regarded as using 3kW continuously, even though rarely switched on all the time. A continuous 3kW load would seriously reduce a ring circuit's capacity, so immersion heaters must have their own radial circuits.

The circuit needs to be run in 2.5mm² two-core-and-earth cable protected by a 15amp fuse. Each element must have a double-pole isolating switch mounted near the cylinder; the switch should be marked 'WATER HEATER' and have a neon indicator **(1)**. A 2.5mm² heat-resistant flexible cord runs from the switch to the immersion heater.

If the cylinder is situated in a bathroom, the switch must be outside zones 0 to 2. If this precludes an ordinary water-heater switch, fit a 20amp ceiling-mounted pull-switch with a mechanical ON/OFF indicator.

Wiring side-entry heaters

For simplicity use two switches, one for each heater and marked accordingly.

Wiring the switches

Fix the two mounting boxes to the wall, feed a circuit cable to each, and wire them in the same way. Strip and prepare the wires, then connect them to the 'Mains' terminals – brown to L, blue to N. Sheathe the earth wire in a green-and-yellow sleeve and fix it to the common earth terminal **(2)**.

Prepare a heat-resistant flex for each switch. At each one, connect the green-and-yellow earth wire to the common earth terminal and the other wires to the 'Load' terminals – brown to L, and blue to N **(2)**. Then tighten the flex clamps and screw on the faceplates.

Wiring the heaters

The flex from the upper switch goes to the top heater, and the flex from the lower switch to the bottom one. At each heater, feed the flex through the hole in the cap and prepare the wires.

Connect the brown wire to one of the terminals on the thermostat (the other one is already connected to the wire running to an L terminal on the heating element). Connect the blue wire to the N terminal, and the green-and-yellow wire to the E terminal **(3)**. Then replace the caps on the terminal boxes.

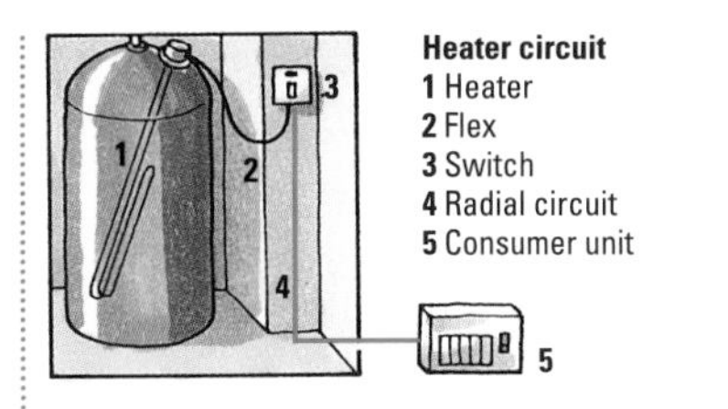

Heater circuit
1 Heater
2 Flex
3 Switch
4 Radial circuit
5 Consumer unit

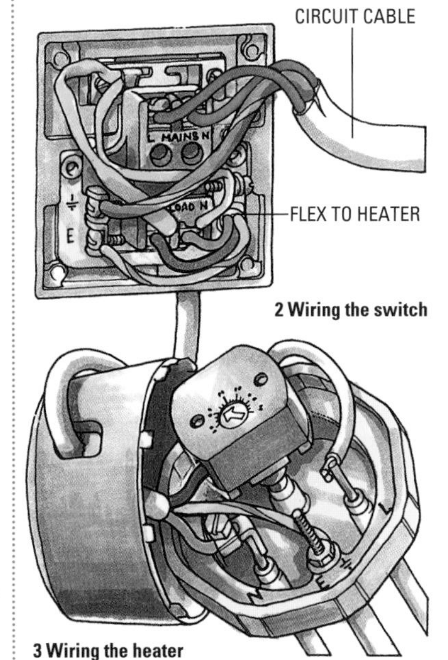

2 Wiring the switch

3 Wiring the heater

• RCD protection
When installing any electrical appliance in a bathroom, the circuit should be protected by a 30 milliamp RCD.

1 A 20amp switch for an immersion heater

Running the cable

Run the circuit cables from the cylinder cupboard to the fuse board; then, with the power switched off, connect the cable from the upper heater to a spare fuseway in the consumer unit. Although the consumer unit is switched off, the cable between the main switch and the meter will remain live – so take special care. Wire the other cable to its own switchfuse unit – or to your storage-heater consumer unit, if you have one – ready for connection to the Economy 7 time switch. Make the connections as described for a cooker circuit.

Dual-element heaters

Wire the immersion-heater circuit as described above, but feed the flex from both switches into the cap on the heater. Connect the brown wire from the upper switch to the L2 terminal on the one thermostat, and the other brown wire to the L1 terminal on the second thermostat **(4)**. Connect the blue wires to their respective neutral terminals **(4)**. Connect both earth wires to the E terminal.

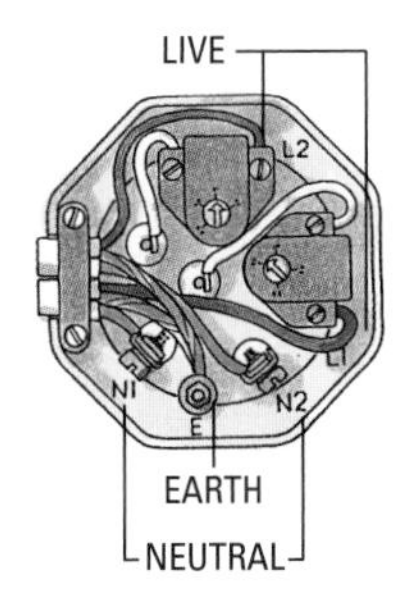

4 Make sure your heater is fitted with two thermostats, as shown.

SEE ALSO: Economy 6, Building Regulations 8, Zones for bathrooms 11, Switching off 16, Consumer units 18, 38, Cooker circuit 37–8, Circuit lengths 66

Storage heaters

The heart of a storage heater is a heat-retaining core, or block, which houses heating elements that are supplied with electricity during the off-peak night-time hours, to take advantage of the cheap Economy 7 tariff. The storage core is insulated in such a way that it will give off heat gradually during the day. Heat emission is controlled in various ways.

With the earliest storage heaters it was not possible to control the rate of heat emission, and towards the end of the day emission tended to diminish. This is no longer a problem. Modern heaters have dampers to regulate the flow of air through the core and control the rate of heat loss. Some heaters have dampers that are controlled automatically by circuits that monitor the air temperature in the room.

Research has shown that a cold day is usually preceded by a proportionally cold night – and the more sophisticated storage heaters are designed to make use of this fact by storing just the right amount of heat during the night to meet the needs of the following day.

Fan-assisted storage heaters have a similar heat-retaining core, which is efficiently insulated to reduce heat loss to an absolute minimum. When the fan is switched on, it draws air into the heater, to be warmed before flowing out into the room. Apart from a very small amount of radiant heat through the casing, heat emission occurs only when required, particularly if the fan is controlled thermostatically.

Storage heaters vary in size. Ratings of ones without fans range from 1.2kW to 3.375kW, and fan-assisted models are rated even higher (up to 6kW). A large area requires a heater with a big heat-retaining core able to store enough heat to warm it; and since cheap-rate power is supplied for only a few hours, a large core needs more powerful elements to charge it completely.

When you install storage heaters, you have to assemble them yourself. Follow the manufacturer's instructions exactly, and handle the heating elements and insulation with care. Make sure slim heaters are fixed to the walls securely – but leave a 75mm (3in) gap all round, so the air can circulate. If possible, use fibre wallplugs for the fixings, as plastic ones may be softened by the heat.

Don't dry clothes on a storage heater; this practice is likely to make a fusible link in the unit melt. Never assemble or dismantle old storage heaters – they may contain asbestos.

Storage-heater circuits

Unlike other kinds of electrical heating, all the storage heaters in a house are usually switched on at the same time – a procedure that would overload a ring circuit. You therefore have to provide an individual radial circuit for each heater. A separate consumer unit is installed to cope with the off-peak load.

It's wise to choose a consumer unit that is not only large enough to take all the heater circuits but has spare fuse-ways for possible additional heaters in the future. Make sure there is an extra fuseway to take the immersion-heater circuit, so your water can be heated at the off-peak rate, too. A circuit for an ordinary storage heater up to 3.375kW should be wired with 2.5mm^2 two-core-and-earth cable, with a 15amp circuit fuse or 16amp MCB.

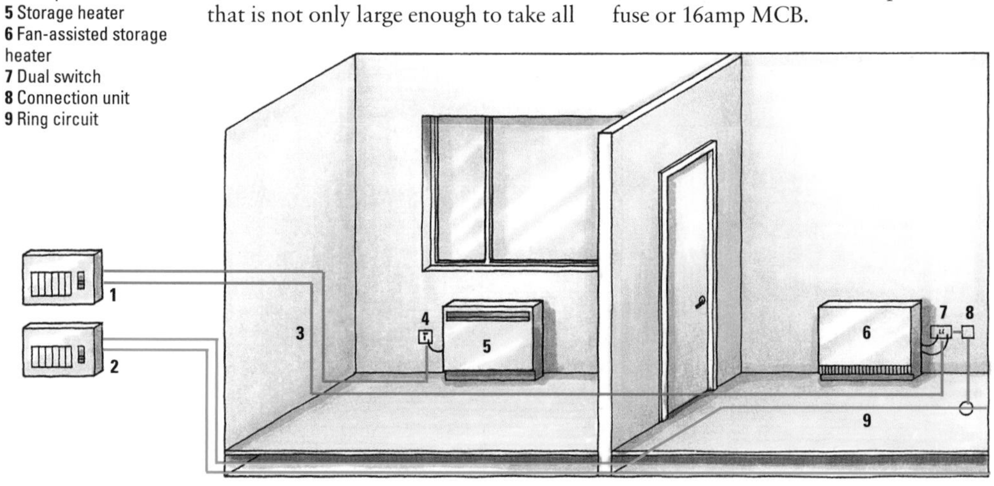

Storage-heater circuits
1 Off-peak consumer unit
2 Day-time consumer unit
3 Radial circuits to heaters
4 20amp switch
5 Storage heater
6 Fan-assisted storage heater
7 Dual switch
8 Connection unit
9 Ring circuit

Because an Economy 7 storage-heater system uses cheap-rate power, a special meter is needed, to register the number of units consumed during the night-time and daytime separately. You will also need a time switch to connect the various circuits at the appropriate time.

This equipment is supplied by the electricity company. It's best to contact them for advice as soon as possible if you plan to have storage heaters. At the same time, check that your present electrical installation is safe, especially the provision for earthing – otherwise the company may refuse to connect the new circuits.

Storage-heater outlets

The circuit cable for an ordinary storage heater should terminate at a 20amp double-pole switch with a flex outlet **(1)** that fits into a standard plastic or metal mounting box. A three-core heat-resistant flex connects the switch to the storage heater.

A fan-assisted heater needs a more complex circuit. The heating elements are supplied from a straightforward radial circuit using 4mm^2 cable, but the fan requires its own circuit for daytime use. Take a spur from a ring circuit to a fused connection unit that has a 3amp fuse, and run a 1.5mm^2 two-core-and-earth cable from the unit for the fan.

The heater and fan circuits both terminate at a special dual switch **(2)** where fan and heater can be isolated simultaneously. Two lengths of heat-resistant flex run from the switch: one to the heater, the other to the fan. A dual switch can be surface-mounted or flush-mounted.

1 Double-pole switch for a storage heater

2 Dual switch

SEE ALSO: Building Regulations 8, Running a spur 31, Fused connection units 34

Wiring storage heaters

For ordinary storage heaters, mount a 20amp switch close to where you are planning to stand each heater. Run a single length of 2.5mm² two-core-and-earth cable from each switch to the site of the new consumer unit, taking the most economical route.

Feed a cable into the mounting box of each switch, then strip and prepare the wires and connect them up to the 'Mains' terminals: brown to L, blue to N. Sleeve the earth wire and connect it to the E terminal **(1)**.

Pass the flex from each heater through the outlet hole in the faceplate of its switch. Strip and prepare the wires, then connect them to the 'Load' terminals: brown to L, blue to N, and the green-and-yellow earth wire to E **(1)**. Tighten the cord clamp and fix the switch into its mounting box.

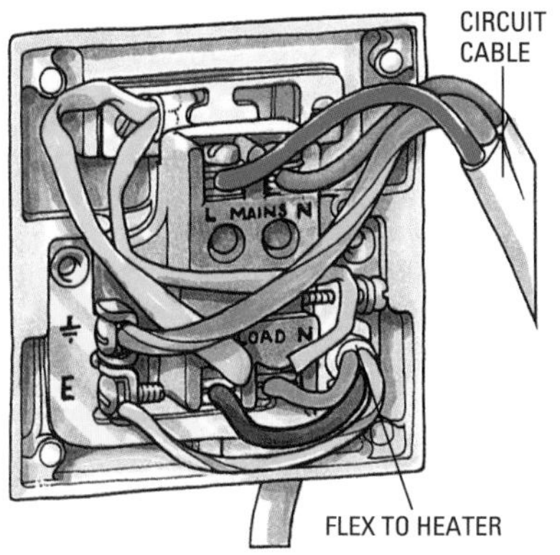

1 Wiring a 20amp switch for a storage heater

Wiring fan-assisted heaters

When you wire a fan-assisted heater, mount a dual switch nearby and from its 'heater' side **(2)** run a 4mm² two-core-and-earth cable to the consumer unit.

Mount a fused connection unit near the switch and run a short length of 1.5mm² two-core-and-earth cable between the two, connecting to the 'Load' side of the connection unit and the 'Mains' side of the fan switch **(2)**.

Run a spur of 2.5mm² two-core-and-earth cable from the 'Mains' terminals on the connection unit **(2)** to either a junction box or a socket outlet on the nearest ring circuit.

Feed the fan and heater flex into the outlets in the faceplate of the dual switch and strip and prepare the wires. Connect each flex to its own part of the switch, which is clearly labelled **(2)**.

Tighten the cord clamps and screw the switch to its box.

For ordinary storage heaters, fit a 15amp cartridge fuse or 16amp MCB for each heater circuit, and a similar fuse or MCB for an immersion-heater circuit if required. Mount the storage-heater consumer unit on an exterior-grade plywood board 9mm (⅜in) thick. Even if there's ample room for it, don't mount it on the electricity company's meter board.

Screw your board to the wall, using plastic or ceramic insulators to space it away, so that damp won't penetrate it. Get the insulators when you buy the consumer unit. Position the board close to the meter, to keep the meter leads as short as possible. Screw the consumer unit to the board; run the circuit cables from the heaters into it one at a time; then prepare the wires for connection.

Each circuit is wired, in the same way, to a separate fuseway: the brown wire to the terminal on the fuseway, the blue one to the neutral block, and the earth wire to the earth block after sheathing it with a green-and-yellow sleeve.

Use 16mm² single-core cable for the meter leads. They must be insulated in brown for the live conductor, and blue for the neutral. Feed the leads into the consumer unit and connect them to their terminals – brown to L, blue to N – on the main isolating switch.

Next, connect a length of green-and-yellow 16mm² single-core cable to the earth block. Connect the other end to the consumer's earth terminal, and a further length of the same-size cable to the same earth terminal – this will be connected to the electricity company's earth by their representative.

Fit MCBs, or clip a fuse into each of the fuse carriers and insert the carriers into their fuseways. Label all of the circuits clearly, so that in future you can tell which heater each one supplies.

Fit the cover on the consumer unit and test the circuits (see right). Then get your Building Control Office to send an inspector to make a final test before you ask the electricity company to connect the unit to the meter and earth. Don't make these connections yourself.

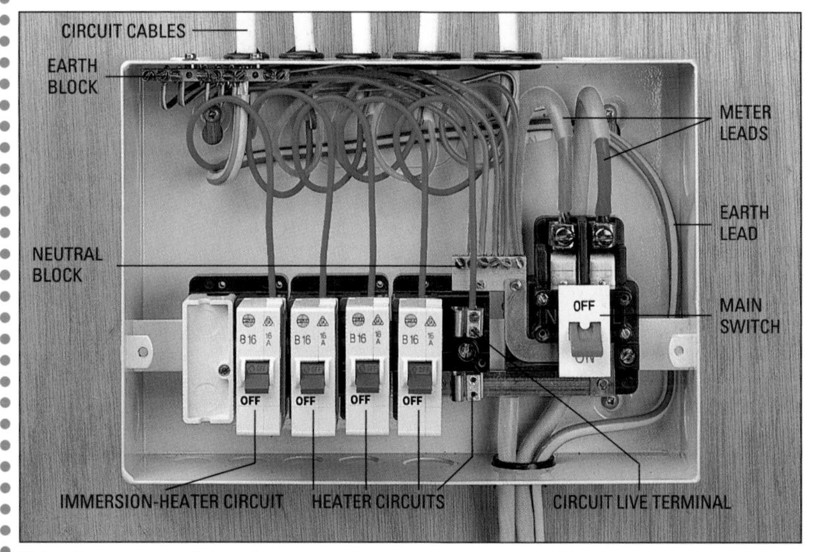

Wiring the consumer unit for ordinary storage heaters

• Fuses and MCBs for fan-assisted heaters
You will need to fit a 30amp circuit fuse or 32amp MCB for each fan-assisted storage heater (see CIRCUITS: MAXIMUM LENGTHS).

• Testing the circuits
Carry out the following tests before asking the electricity company to connect your new meter leads.

At your new consumer unit, switch all the MCBs and the main switch on. Turn all the heater switches off. Put one probe of a continuity tester on your neutral meter lead and the other probe on the earth wire. The tester's indicator should not illuminate.

Next, put the probes between the new live and neutral meter leads. Again the tester's indicator should not illuminate.

Now, leaving the probes on the neutral and live meter leads, turn each heater switch on in turn. Each time, the indicator should illuminate. If any of the tests fail, check your wiring or seek expert advice.

The BCO inspector may want to repeat these tests, using a 500V instrument.

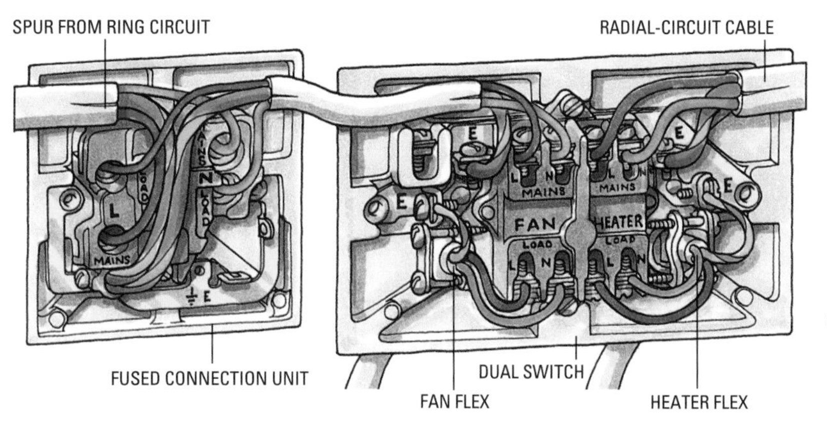

2 Wiring a dual switch
Connect a fused connection unit to the dual switch.

SEE ALSO: Stripping flex 13, Fuses/MCBs 19, Cables 22, Stripping cable 22, Running cable 23–5, Running a spur 31, Circuit lengths 66, Final tests 67

Door bells and chimes

Whether you choose a door bell, a buzzer or a set of chimes, there are no practical differences that affect the way they are installed.

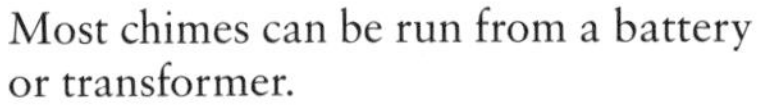

Door bells

Most door bells are of the 'trembler' type. When electricity is supplied to the bell – that is when someone presses the button at the door – it activates an electromagnet, which causes a striker to hit the bell. But as the striker moves to the bell it breaks a contact, cutting off power to the magnet – so the striker swings back, makes contact again and repeats the process, going on for as long as the button is depressed. This type of bell can be operated by battery or (if it is an AC bell) by a mains transformer, which may be situated inside the unit or mounted separately.

Buzzers

A buzzer operates on exactly the same principle as a trembler bell, but in a buzzer the striker hits the magnet itself instead of a bell.

Chimes

A set of ordinary door chimes has two tubes or bars tuned to different notes. Between them is a solenoid, containing a spring-loaded plunger which acts like the trembler striker described above. Most chimes can be run from a battery or transformer.

Chimes
A set of chimes has two tubes, each tuned to a different note.

Bell pushes

Pressing a bell push completes the circuit that supplies power to the bell. The bell push is in effect a switch that is operated by holding it in the 'on' position. Inside it are two contacts, to which the circuit wires are connected **(1)**. One contact is spring-loaded, touching the other when the button is depressed, to complete the circuit, and then springing back again when the button is released.

Illuminated bell pushes incorporate a tiny bulb, which enables you to see the bell push in the dark. These have to be operated from a mains transformer – as the power to the bulb, although only a trickle, is on continuously and would soon drain a battery. Luminous types glow at night without a power supply.

Wireless bell pushes

To remove the need for wiring, you can use a bell push that sends a radio signal to its bell. The bell can be moved around the house; and you can add a second bell, if required.

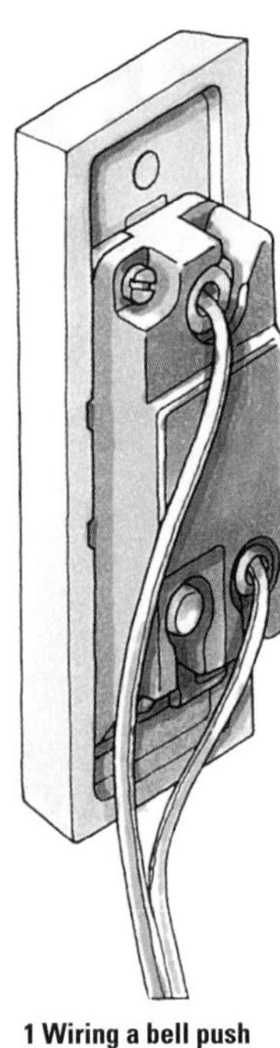

1 Wiring a bell push

Batteries or transformer?

Some bells and chimes house batteries inside the casing, while other types incorporate a built-in transformer that reduces the 230V mains electricity to the very low voltages needed for this kind of equipment. For many door bells or chimes you can use either method. Most of them take either two or four 1½V batteries, but some need a 4½V battery that is housed separately.

The transformers sold for use with door-bell systems have three low-voltage tappings (3V, 5V and 8V), to cater for various needs. Generally 3V and 5V connections are adequate for bells or buzzers; the 8V tapping is suitable for many sets of chimes.

However, some chimes require a higher voltage, and for these you will need a transformer with 4V, 8V and 12V tappings. A bell transformer must be designed in such a way that the full mains voltage cannot cross over to the low-voltage wiring.

Circuit wiring

The battery, bell push and bell are all connected by two-core insulated 'bell wire'. This fine wire is usually surface-run, fixed with small staples; but it can be run under floors and in cupboards, too. Bell wire is also used for connecting up a bell and bell push to a transformer.

Connect a BS 3535 Part 2 double-insulated transformer to a junction box or ceiling rose on a lighting circuit with 1mm² two-core-and-earth cable. As no earth is required for a double-insulated transformer, cut and tape back the earth wire at the transformer end.

Alternatively, run a spur from a ring circuit in 2.5mm² two-core-and-earth cable to an unswitched fused connection unit, fitted with a 3amp fuse; then run a 1mm² two-core-and-earth cable from the connection unit to the transformer's 'Mains' terminals.

INSTALLING A SYSTEM

The bell itself can be installed in any convenient position, so long as it isn't over a source of heat. The entrance hall is usually best, as a bell there can be heard in most parts of the house.

Keep the bell-wire runs as short as possible, especially for a battery-operated bell. With a mains-powered bell you will want to avoid long and costly runs of cable – so position the transformer where it can be wired simply. A cupboard under the stairs is usually a good place.

Drill a small hole in the doorframe and pass the bell wire through to the outside. Fix the conductors to the terminals of the bell push, then screw it over the hole.

If the battery is housed in the bell casing, there will be two terminals for attaching the other ends of the wires. Either wire can go to either terminal. If the battery is separate from the bell, run the bell wire from the push to the bell. Separate the conductors, cut one of them and join each cut end to a bell terminal. Run the wire on to the battery and attach it to the terminals **(1)**.

Wiring to a transformer

If you are wiring to a transformer, proceed as above but connect the bell wire to whichever two of the three terminals combine to provide you with the necessary voltage **(2)**. Some bells and chimes require separate lengths of bell wire, one from the bell push and another from the transformer. Fix the wires to terminals in the bell housing, following manufacturer's instructions.

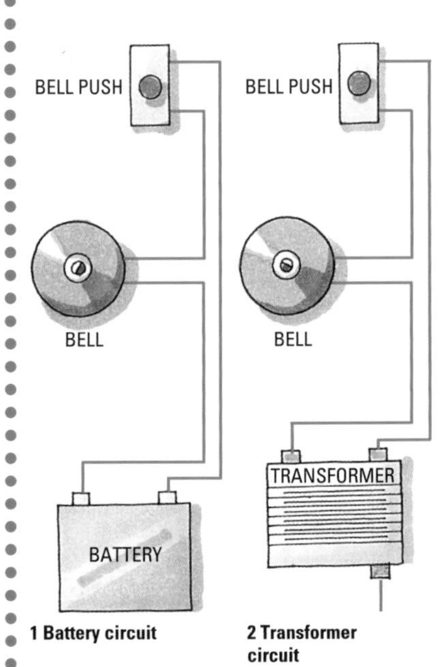

1 Battery circuit **2 Transformer circuit**

SEE ALSO: Consumer units 18, 38, Running cable 23–5, Running a spur 31, Connecting to a light circuit 53

Wiring aerial sockets

Many people operate only one television set from an aerial mounted on the roof of their home, and rely on portable aerials for any additional sets. You can improve reception by extending the main aerial with additional sockets and, at the same time, provide for viewing a video-cassette recorder from any of your television sets.

One convenient arrangement is to connect the output socket from your VCR to a double aerial socket, which acts as a 'splitter', diverting the signal to two television sets. Each set will work independently of the other.

If you want to serve even more sets, you will probably have to substitute a multi-output amplifier in place of the splitter in order to boost the signal. An amplifier is wired in a similar way to the splitter socket, but must also be plugged into a 13amp socket.

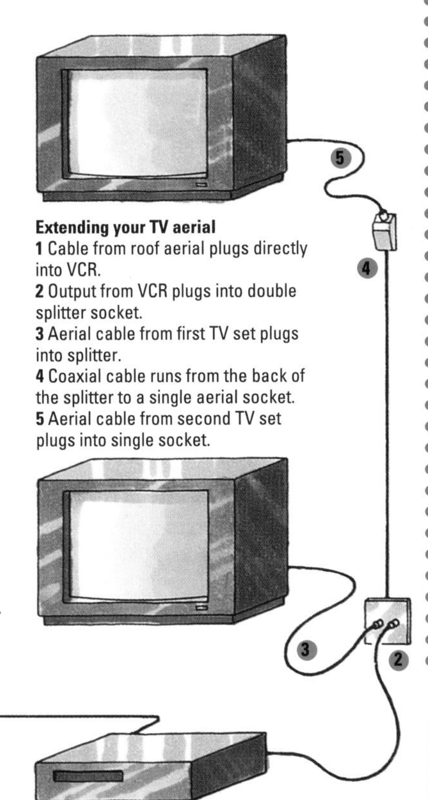

Extending your TV aerial
1 Cable from roof aerial plugs directly into VCR.
2 Output from VCR plugs into double splitter socket.
3 Aerial cable from first TV set plugs into splitter.
4 Coaxial cable runs from the back of the splitter to a single aerial socket.
5 Aerial cable from second TV set plugs into single socket.

Cable and equipment

Aerial sockets are wired with coaxial cable. This consists of a single-core solid-copper conductor, insulated with polythene, which is surrounded by a braided conductor woven from many fine copper strands then sheathed in white, brown or black PVC. Most electrical suppliers stock the required 75 Ohm cable **(1)**. Coaxial cable is either wired directly into the back of aerial sockets or fitted with special plugs **(2)** for insertion into the sockets.

You can buy single and double aerial sockets with square faceplates **(3)** for attaching to standard plastic or metal mounting boxes. There are also small surface-mounted sockets **(4)** suitable for screwing to skirting boards.

A double socket can serve to split the incoming signal to two television sets – but if the signal is weak, you may find reception is not satisfactory. In which case, either install a signal amplifier or use a switched splitter **(5)**, which allows you to divert the full-strength signal to one set or the other at will.

It is simplest to install only 'female' sockets – ones with holes that accept 'male' coaxial plugs – and fit male plugs on all your aerial-extension cables.

Cable, plugs and sockets
1 Coaxial cable **2** Coaxial plug **3** Double-socket faceplate **4** Surface-mounted socket **5** Switched splitter socket

CONNECTING COAXIAL PLUGS

Slide the plug's locking ring **(1)** onto the coaxial cable and strip about 25mm (1in) of the sheathing – taking care not to sever the copper strands beneath. Slide the cable gripper **(2)** onto the end of the sheathing. Then unravel the copper strands and fold them down over the gripper. Cut off excess strands, leaving enough copper to cover the gripper.

Strip all but about 3mm (⅛in) of the polythene insulation **(3)** to reveal the single-core conductor **(4)**. Bend a slight kink in the conductor and insert it in the plug pin **(5)**. Ideally you should secure the conductor with a touch of solder on the tip of the pin, though kinking the conductor usually provides sufficient grip inside the hollow pin. Finally, slide the plug body **(6)** over the whole assembly and secure it with the locking ring.

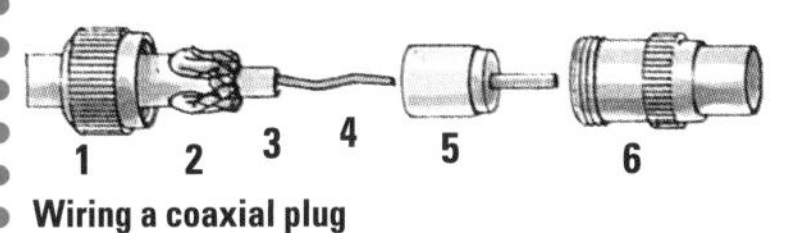

Wiring a coaxial plug

Running coaxial cable

Fit mounting boxes or screw sockets to the skirting in convenient positions for your VCR and television sets, then cut suitable lengths of coaxial cable to run from socket to socket. Although it's quite safe to leave coaxial cable as temporary unfixed 'leads', you can conceal cable runs under floorboards and inside wall cavities – in the same way as you would mains cable. Avoid taking the cable around tight bends.

Wiring the splitter

Prepare a length of coaxial cable to connect the back of the double socket that is to act as a splitter to the back of the single remote socket. Strip about 50mm (2in) of sheathing from the splitter end of the cable and fold back the braided copper strands. Strip about 32mm (1¼in) of insulation from the single-core conductor, then pass the conductor through both terminals **(6)** and tighten the terminal screws. Fold back the braided copper and trap it, along with the cable, under the metal clamp **(6)**. Trim off excess copper strands, then screw the faceplate to the mounting box. The remote socket is wired in a similar way.

● Digital TV
The transmission of analogue TV signals is being phased out in favour of digital signals. Set-top converters are available to allow digital reception on existing equipment. Aerials for analogue reception are suitable for digital reception.

● Satellite dishes
Installing your own dish aerial is a simple DIY project, but it involves altering the direction of the aerial to obtain the strongest signal. This is probably best left to the TV supplier or satellite station, who usually offer free installation as part of the package.

6 Wiring a double socket as a splitter

SEE ALSO: Running cable 23–5, Mounting boxes 28–9

Telephone extensions

Although a telephone company, such as British Telecom or Mercury, must be employed to install the master socket that is connected to the incoming network cable, you are permitted to install extension cables and sockets yourself.

All the necessary equipment is available from DIY outlets or from one of the telephone company's own shops.

You can install as many telephone extension sockets as you want, so long as the total 'Ringer Equivalence Number' (REN) in your house or flat doesn't exceed four. A telephone is normally allocated an REN of one – but it is advisable to check this before you decide which equipment to purchase. Telephones are made with either 'tone' or 'pulse' dialling, and modern phones can be switched from one to the other. However, the type of dialling does not affect the wiring of extension sockets.

Sockets and accessories
1 Single-socket faceplate
2 Surface-mounted socket
3 Socket doubler
4 Converter plug
5 British Telecom Linebox
6 Insertion tool

Telephone sockets

Single and double sockets designed to accept the small rectangular telephone plugs are made in the form of square faceplates **(1)** that fit standard electrical metal and plastic mounting boxes. Compact surface-mounted sockets are also available **(2)**.

To operate two telephones or a telephone and an answering machine from a single socket without additional wiring, simply plug in a 'socket doubler' **(3)**.

You can run an extension from any master socket by means of a converter plug **(4)**, which usually comes complete with several metres of cable. Another option is to wire your extension cable directly into a British Telecom Linebox **(5)**, which has a removable cover to give customer access without disturbing the telephone company's wiring.

TELEPHONE CABLE

Telephones, including extensions, are wired with extra-low-voltage cable.

Telephone cable usually comprises six colour-coded conductors sheathed in PVC. However, four-core cable is often sold for running domestic telephone extensions, and is perfectly adequate provided you match the colour-coded conductors to any existing wiring (see chart below).

Socket terminals are numbered 1 to 6. Always match the same colour coding to the same number terminal in each socket. If you are using four-core cable, ignore terminals 1 and 6.

Number	Colour coding
Terminal 1	Green with white rings.
Terminal 2	Blue with white rings.
Terminal 3	Orange with white rings.
Terminal 4	White with orange rings.
Terminal 5	White with blue rings.
Terminal 6	White with green rings.

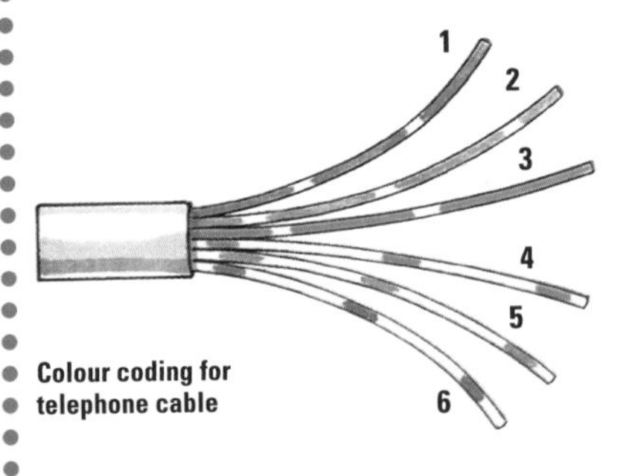

Colour coding for telephone cable

1 Connecting cable to blade terminals
In some sockets, there will be two identical wires per terminal.

Running the circuit

Fix the extension sockets where most convenient, and run a length of cable from the existing master socket to each of the extension sockets. The cable can be pinned to the top of skirting boards or along picture rails and doorframes, using small plastic cable clips.

Alternatively, you can conceal the cable under the floorboards or within walls, provided that they do not follow exactly the same route used for mains wiring. Both for safety and in order to avoid interference on the line, maintain a minimum of 75mm (3in) between the telephone cable and any mains cables.

At each of the sockets, feed a loop of cable into the mounting box, ready for connecting to the terminals.

Connecting to the sockets

At each socket, cut the loop of cable and strip the sheathing to expose the colour-coded conductors, then separate the conductors and connect them to the appropriate numbered terminals.

Telephone-socket terminals usually comprise two opposing brass blades that cut into the cable's insulation and make contact with the wire core as the conductor is forced between them with a special insertion tool. Lay the insulated conductor across its terminal, and press it firmly to the base of the terminal **(1)**. Trim the end of the wire.

Other sockets are made with screw terminals, similar to those found in 13amp plugs. Strip about 6mm (¼in) of insulation from the end of each of the conductors, then insert the wire into the terminal and tighten the screw **(2)**.

Sometimes, plastic cable ties are provided to secure the cable inside the socket, in order to prevent strain on the actual connections.

2 Insert wires into screw terminals

Wiring the master socket

Plugging a converter plug into the master socket will connect all your extensions to the telephone company network. To connect cable to a British Telecom Linebox, remove the front cover **(3)** and use the insertion tool to introduce the conductors into the bladed terminals, as described left.

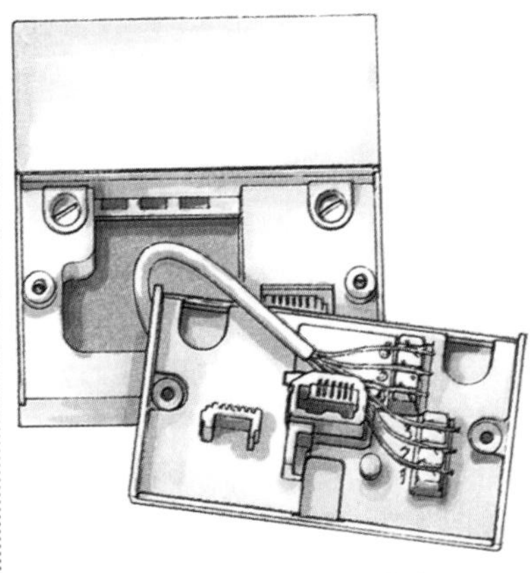

3 Connecting to a British Telecom Linebox

SEE ALSO: Running cable 23–5, Mounting boxes 28–9

An office at home

Working from home has become a practical option for a great many people. And even those who commute to the workplace usually need somewhere at home where they can catch up with extra work and sort out personal accounts. Homework and hobbies put the younger members of the family in a similar position.

Increasingly, these activities are centred on a computer and a network of electronic equipment. Whether you make do with a corner of the dining table or have the luxury of a dedicated workspace, some planning – and perhaps new wiring – will avoid a tangle of trailing flexes and overloaded socket outlets.

To assess the number and positions of socket outlets, first plan your office layout to make the best use of the space available. Think about where your desk or worktable should be placed. You will probably want to take advantage of natural light – but before you make any permanent alterations, try out the position of your computer monitor to avoid distracting reflections from windows and fixed lighting.

A worktop for your computer

Most people can work comfortably on a worktop that is 700mm (2ft 4in) from the floor. Ideally a computer keyboard should be slightly lower – which is why ready-made computer work stations are usually made with a slide-out work surface that can be stowed beneath the monitor. If your children are likely to use the same workspace, get a chair that is adjustable in height.

A fixed worktop needs to be at least 600mm (2ft) deep to provide enough room for the average computer and keyboard. But you may need a worktop 750mm (2ft 6in) deep to accommodate larger equipment, unless you can build an L-shape unit that allows the monitor to be tucked into the corner.

You will need extra worktop space for papers and reference books, plus shelves or drawers to store items such as stationery and computer discs.

ANCILLARY EQUIPMENT

Even the most sophisticated computer is of limited use without some ancillary equipment.

You'll need a printer for correspondence and accounts, and to print out e-mails and information downloaded from the Internet. There are plenty of inexpensive colour printers designed for the home user; and you may want a scanner for putting your own photos and graphics onto the computer, which allows you to manipulate and recompose images then print them to a very high quality.

Your computer won't come ready equipped with all of the different disc drives used for copying information onto back-up discs, so you may have to supplement your existing hardware.

And as time goes by, you may well require extra memory for data storage, which might mean a second hard drive for the computer. Then there's your modem for Internet and e-mail access. And perhaps a CD or DVD writer. The list goes on and on.

The cable jungle

Each new piece of equipment needs a power supply and a connection to the computer – which is why so many home offices end up with a tangle of wires and overloaded sockets.

To reduce the number of cables, you could get a computer with an internal modem and disc drives; or have these items – and if need be, an extra hard disc – installed in your present machine. Another option is to buy stand-alone equipment powered from the computer itself, instead of from sockets.

Whenever possible, buy accessories that are connected via a USB (Universal Serial Bus), as these can be swapped around without having to switch off the computer. Better still, connect all your equipment to a USB hub plugged into the back of the machine.

Whatever measures you take, it's impossible, using current equipment, to eliminate cables altogether. Having enough sockets positioned where needed is therefore still part of the equation.

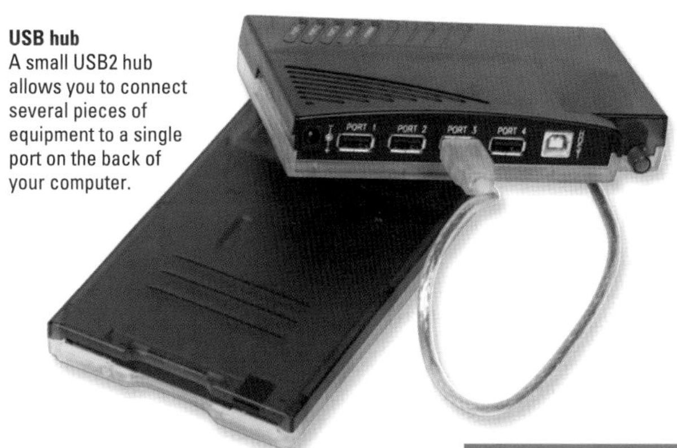

USB hub
A small USB2 hub allows you to connect several pieces of equipment to a single port on the back of your computer.

Work station
Purpose-made unit with a sliding work surface for a keyboard.

Lighting your office
Use dedicated task lighting to illuminate the work area without creating distracting reflections. A portable desk lamp is one option, or you could install a small spot light or downlighter above the workstation. A dimmer switch that controls the room lighting will allow you to set the optimum level of background illumination.

SEE ALSO: Providing extra sockets 46, Dimmer switches 51, 52, Adding wall lights 55, Low-voltage lighting 56–7

Providing extra sockets

Extending the ring circuit is the safest and most efficient means of incorporating new socket outlets to power all the equipment you are likely to need in a home office. You will then be able to plug in as many appliances as you wish, including electric heaters, without fear of overloading the circuit. In addition, it will reduce the risk of a plug being pulled out accidentally, which could result in the loss of irreplaceable data.

Surge protection
Sensitive electronic components in a computer can be damaged by voltage 'spikes' – short-duration peaks of high voltage. You can buy special plugs and trailing sockets fitted with surge suppressors designed to protect vulnerable equipment.

Extending a ring circuit

Most householders just don't have the space to dedicate a room exclusively to working from home – but even if your study has to double as a spare bedroom from time to time, adding a number of double or triple sockets will be time and money well spent.

By far the best method is to break into the ring circuit and connect a new length of cable, either to the existing sockets or by means of junction boxes. Be generous with your new sockets: they are relatively cheap to install, and you can never have too many.

You'll probably find that one or two sockets are needed at skirting level, and two or three more at desktop height. This arrangement will provide you with the most direct route for connecting floor-standing and desktop equipment, without having to extend flexible cords. Where possible, rewire plugs – making the flex as short as practicable.

Label all plugs powering vulnerable equipment, such as your computer. This simple precaution will reduce the risk of the equipment being unplugged inadvertently by another member of your family who wants to use the socket for another appliance.

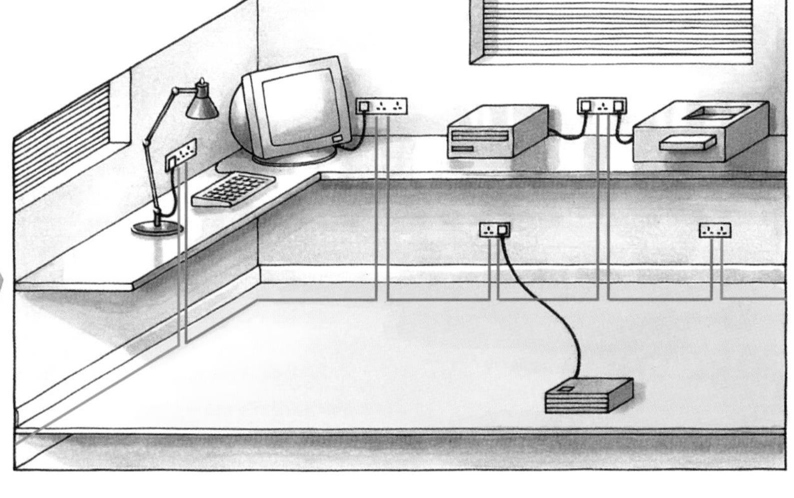

Adding sockets
Extend the ring circuit to add extra sockets at skirting level and to provide a source of power at desktop height.

Telephones and modems

In order to access the Internet, you will need a telephone socket near your computer to connect a modem. To utilize a single telephone socket, you can plug your phone and modem leads into a socket doubler (if your computer does not have a built-in phone socket) – but if you or your family will be 'surfing the Net' a lot of the time, then it may be worth installing a second phone line that is permanently connected to your computer. British Telecom or a similar telephone company can advise you on alternative connection systems, such as ISDN or ADSL. These allow for faster transmission of data and can be used to provide simultaneous use of a telephone and a computer modem.

Many people use an answerphone to pick up messages when they're not at home or are unavailable. These need to be connected to a telephone socket, and to a power supply via a 13amp plug. It is worth labelling this plug, in order to avoid accidental disconnection and the subsequent inconvenience of having to reprogram the unit.

If you don't have a dedicated photocopier, you can equip your home office with a multipurpose printer/copier/fax machine or printer/copier/scanner, which is relatively inexpensive.

Backing up
Develop the habit of making copies of all your important computer files. Copy them onto a disc at the end of each working day, and store one copy somewhere other than in your study.

All back-ups are a waste of time until you need them – then they are priceless!

If extending your ring circuit is not a viable option, there are various ways of connecting more than one appliance to existing sockets outlets.

Trailing sockets

Trailing sockets are made with up to six 13A socket outlets, connected via a short flexible lead to a single plug. With this type of device you can connect your computer and ancillary equipment to a single wall-mounted socket. Look for trailing sockets fitted with surge suppressors (see far left).

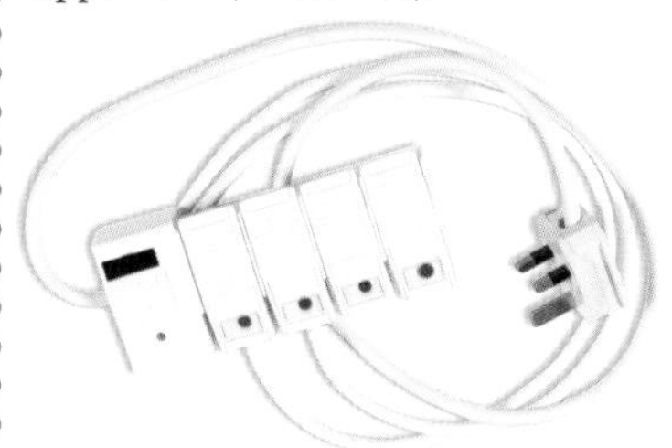

Miniature trailing sockets

There are special trailing sockets made with plugs similar in size to those used to connect a monitor to the back of a computer. They can be screwed to a skirting or the wall behind the desktop.

However, there is one disadvantage. Because the miniature plugs are not fused individually, a fault in any of the appliances connected to the trailing socket will cause the fuse in the 13amp plug to blow, and all of the appliances – including your computer – will be disconnected instantly. To protect vital data, have your computer plugged into its own wall socket and use the trailing socket for ancillary equipment only.

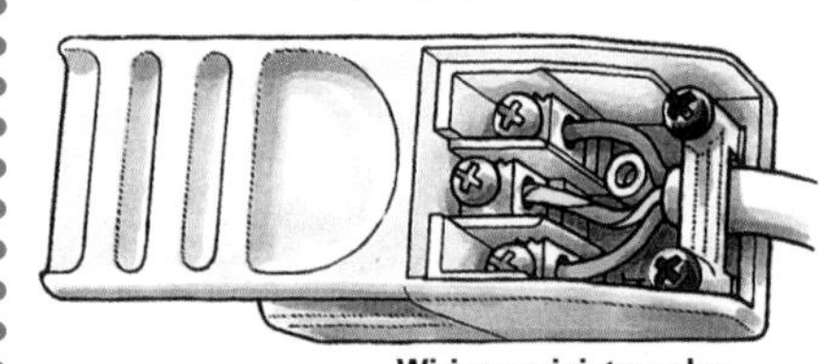

Wiring a miniature plug

Multi-adaptor

You can wire up to four appliances directly to a multi-adaptor, which has a short flex and 13amp plug for connecting to a single wall socket. The flex from each appliance is wired to its own set of terminals inside the adaptor, where it is protected by an individual fuse. Consequently, a fault on a single appliance is less likely to affect other equipment connected to the adaptor.

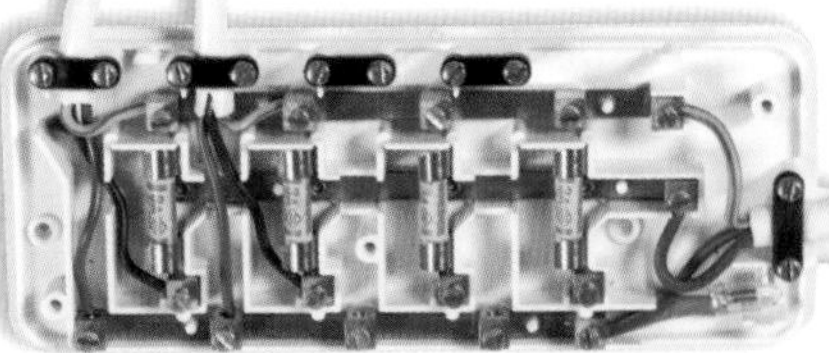

Each appliance is wired to a set of terminals

SEE ALSO: Extending a ring circuit 32, Installing a telephone extension 44

Lighting circuits

An electrically heated shower unit is plumbed into the mains water supply. The flow of water operates a switch to energize an element that heats the water on its way to the shower spray-head. Because there's so little time to heat the flowing water, instantaneous showers use a heavy load – from 6 to 10.8kW. Consequently, an electrically heated shower unit has to have a separate radial circuit, which must be protected by a 30 milliamp RCD.

The circuit cable needs to be 10mm^2 two-core-and-earth. For showers up to 10.3kW, the circuit should be protected by a 45amp MCB or fuse, either in a spare fuseway at the consumer unit or in a separate single-way consumer unit fitted with a 30 milliamp RCD. A 10.8kW shower needs a 50amp MCB. The cable runs directly to the shower unit, where it must be wired according to the manufacturer's instructions.

The shower unit itself has its own on/off switch, but there must also be a separate isolating switch in the circuit. This must not be accessible to anyone using the shower, so you need to install a ceiling-mounted 45amp double-pole pull-switch (a 50amp switch is required for a 10.8kW shower). The switch has to be fitted with an indicator that tells you when the switch is 'on'. Fix the backplate of the switch to the ceiling and, having sheathed the earth wires with a green-and-yellow sleeve, connect them to the E terminal on the switch. Connect the conductors from the consumer unit to the switch's 'Mains' terminal, and those of the cable to the shower to the 'Load' terminals **(1)**.

The shower unit and all metal pipes and fittings must be bonded to earth.

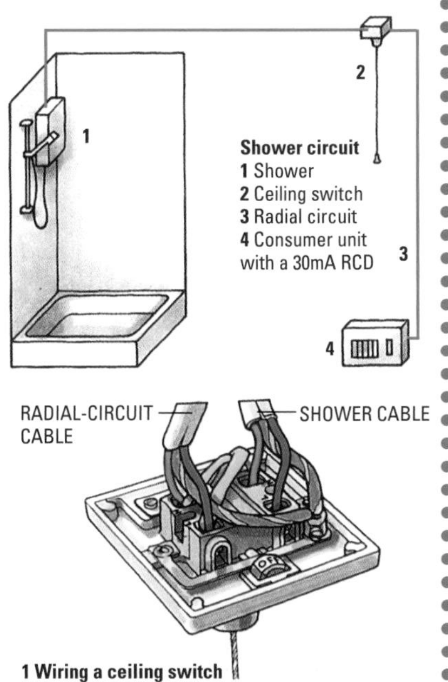

1 Wiring a ceiling switch

Every lighting system needs a feed cable to supply power to the various lighting points, and a switch that can interrupt the supply to each point. There are two ways of meeting these requirements in your home: the junction-box system and the loop-in system. Your house may be wired with either one – though it's quite likely that there will be a combination of the two systems.

The junction-box system

With a junction-box system, a two-core-and-earth feed cable runs from a fuseway in the consumer unit to a series of junction boxes, one for each lighting point. From each junction box a separate cable runs to the light itself, and another runs to its switch.

The loop-in system

With the loop-in system, the ceiling rose takes the place of the junction box. The cable from the consumer unit runs into each rose and out again, then on to the next. The switch cable and the flex to the bulb are connected at the rose.

Combined system

The loop-in system is now more widely used since it entails fewer connections, as well as saving on the cost of junction boxes. However, lights located at some distance from a loop-in circuit are often run from a junction box on the circuit in order to save cable; and lights added after the circuit has been installed are also often wired from junction boxes.

The circuit

Both the junction-box system and the loop-in system are, in effect, multi-outlet radial circuits. The cable runs from the consumer unit, looping in and out of the ceiling roses or junction boxes, and terminates at the last one. Unlike the cable of a ring circuit, it doesn't return to the consumer unit.

Lighting circuits require 1mm^2 or 1.5mm^2 PVC-insulated-and-sheathed two-core-and-earth cable, and each circuit is protected by a 5amp circuit fuse or 6amp MCB. A maximum of eleven 100W bulbs or their equivalent can therefore use the circuit.

In the average two-storey house the usual practice is to have two separate lighting circuits – one for the ground floor and the other for upstairs.

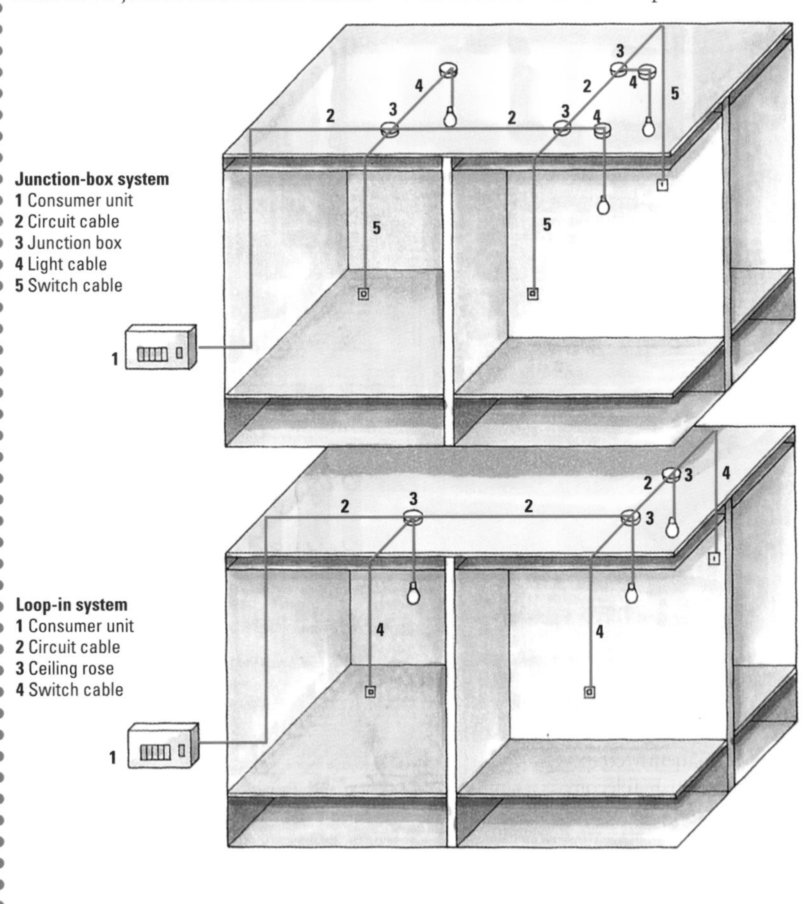

SEE ALSO: Building Regulations 8, Bonding to earth 10, Zones for bathrooms 11, Consumer units 18, Fuses 19, Cables 22, Running cable 23–5, Fixing to ceiling 48, Circuit lengths 66

Identifying connections

● **Old colour coding**
When working on a house built before 2005 you are likely to find that the existing cables are colour-coded black for neutral and red for live. The diagrams on this page show old-style lighting-circuit cables. For new-style circuits, substitute brown for red and substitute blue for black.

Loop-in system

A loop-in ceiling rose has three terminal blocks, arranged in a row. The live (red) wires from the two cut ends of the circuit-feed cable run to the central live block, and the neutral (black) wires run to the neutral block on one side. The earth wires (green-and-yellow) run to a common earth terminal **(1)**.

The live (red) wire from the switch cable is connected to the remaining terminal in the central live block. The electricity runs through this wire to the switch, then back to the ceiling rose via the black 'switch-return' wire, which is connected to the third terminal block in the ceiling rose (the 'switch-wire block'). When the light is 'on', the switch-return wire is live – it is therefore important to identify it by wrapping a piece of red tape round it to distinguish it from the other black wires, which are neutral. The earth wire in the switch cable goes to the common earth terminal **(1)**.

The brown (live) wire from the flex of the pendant light connects to the remaining terminal in the switch block, while the blue wire runs to the neutral block. If three-core flex is used, the green-and-yellow earth wire runs to the common earth terminal **(1)**.

If the circuit-feed cable terminates at the last ceiling rose on the circuit, then only one set of cable conductors is connected **(2)**. The switch cable and the light flex are connected in the same way as those in a normal loop-in rose.

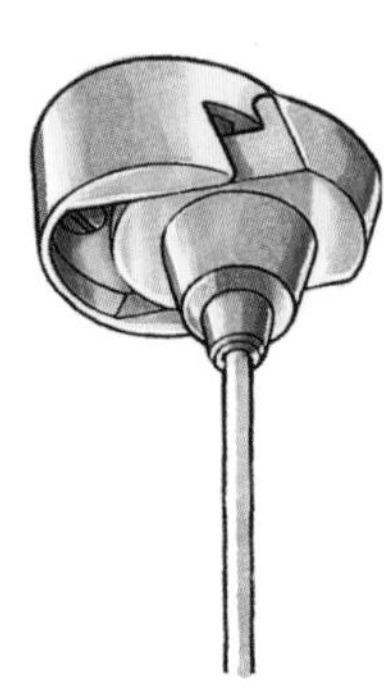

Detachable ceiling roses
If you use a modern detachable ceiling rose, you can slide out the centre section to change the light fitting without disturbing the fixed wiring. This type of rose can support a light fitting weighing up to 5kg (11lb).

Junction-box system

The junction boxes on a lighting circuit normally have four unmarked terminals – for live, neutral, earth and switch connections. The live, neutral and earth wires from the circuit-feed cable go to their respective terminals **(3)**.

The live (red) wire from the cable that runs to the ceiling rose is connected to the switch terminal; the black wire to the neutral terminal; and the green-and-yellow earth wire to the earth terminal **(3)**.

The red wire from the switch cable is connected to the live terminal; the earth wire to the earth terminal; and the black 'return' wire (see above) from the switch goes to the switch terminal **(3)**. This last conductor must be clearly identified with a piece of red tape wrapped round it.

At the ceiling rose, the live cable wire is connected to one of the outer terminal blocks, and the neutral wire to the other one. The central block is left empty. The earth wire goes to the earth terminal **(4)**.

The brown flex wire is connected to the same terminal block as the red cable wire; and the blue flex wire goes to the block holding the black cable wire. If the flex has a green-and-yellow earth wire, it should be connected to the common earth terminal **(4)**.

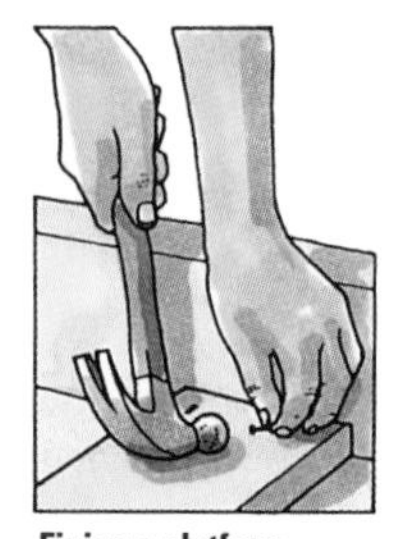

Fixing a platform
Skew-nail a board between the joists to support a ceiling rose.

Replacing a ceiling rose

Turn off the power at the consumer unit and remove the circuit fuse or lock off the MCB. Switching off at the wall is not enough. Unscrew the rose cover and examine the connections. If it's a loop-in rose, identify the switch-return wire with red tape. If there's only one red and one black wire, it's a junction-box system and there will be no switch cable.

Identify the wires inside an old rose. If there are wires running into three terminal blocks, look first for the one with all red wires and no flex wires. This is the live block, containing live circuit-feed wires and a live switch wire. The neutral terminal block takes the black circuit-feed wires and the blue flex wire. The third block will contain the brown flex wire plus a black wire (the switch-return wire), which should be marked with red tape. All earth wires will run to one terminal on the backplate.

Fixing the new rose
Disconnect the wires from the terminals and identify them with tapes. Take down the old backplate. Knock out the entry hole in the new backplate and thread the cables through it; then fix the backplate to the ceiling, if possible using the old screws and fixing points. If the old fixings aren't secure, nail a piece of wood between the joists above the ceiling (see left) and drill a hole through it from below for cable access. Screw the new rose backplate to the wood through the ceiling, and then reconnect all the wires.

Slip the new cover over the pendant flex and connect the flex wires to the terminals in the rose – loop these wires over the rose's support hooks to take the weight off the terminals. Screw the cover onto the backplate, then switch on the power and test the light.

● **No earth wires**
If you uncover an old lighting system that lacks earth wires, reconnect the other wires temporarily and get expert advice on rewiring the circuit.

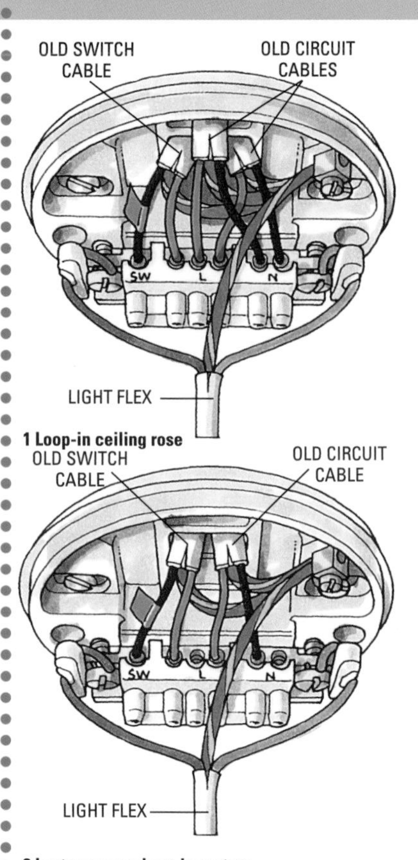

1 Loop-in ceiling rose

2 Last rose on a loop-in system

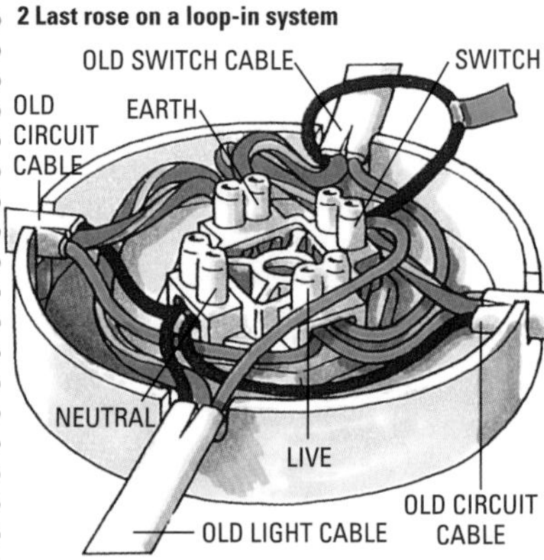

3 Lighting junction box

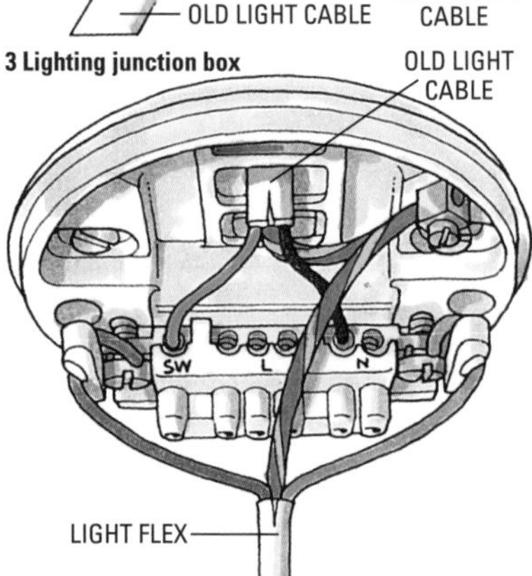

4 Ceiling rose on a junction-box system

SEE ALSO: Lighting circuits 47

Light fittings

There is a vast range of light fittings for the home – but, although they may differ greatly in their appearance, they can be grouped roughly into six basic categories according to their functions.

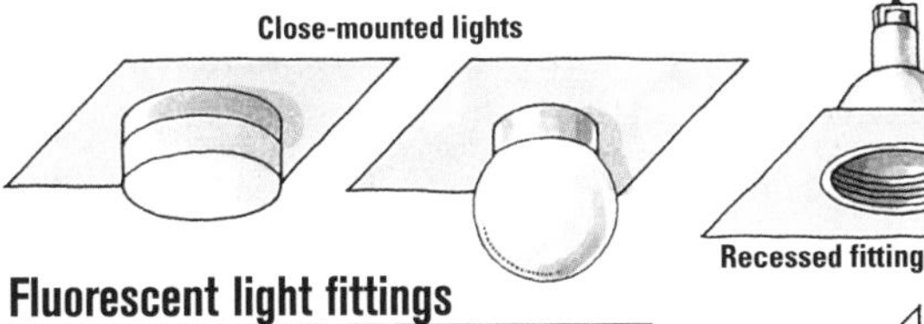

Pendant lights

The pendant light is probably the most common fitting. At its most basic, it consists of a lampholder, with a bulb and usually with some kind of shade, and is suspended from a ceiling rose by a length of flex.

Many decorative pendant lights are designed to take more than one bulb, and they may be much heavier than the simpler ones. Heavy pendant lights should never be attached to a standard plastic ceiling rose. However, they can be connected to a detachable ceiling rose (see opposite).

Close-mounted ceiling lights

A close-mounted fitting is screwed directly to the ceiling, without a ceiling rose, most often by means of a backplate that houses the lampholder or holders. The fitting is usually enclosed by some kind of rigid light-diffuser, also attached to the backplate.

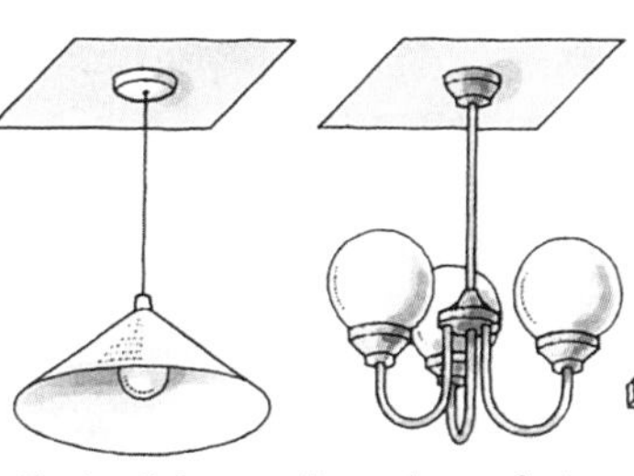

Recessed ceiling lights

The lamp housing itself is recessed into the ceiling void, and the diffuser either lies flush with the ceiling or projects only slightly below it. These discreet light fittings are ideal for rooms with low ceilings; they are often referred to as downlighters.

Track lights

Several individual light fittings can be attached to a metal track, which is screwed to the ceiling or wall. Because a contact runs the length of the track, lights can be fitted anywhere along it.

Wall lights

Light fittings designed for screwing to a wall can be supplied either from the lighting circuit in the ceiling void or from a fused spur off a ring circuit. Various kinds of close-mounted fittings and adjustable spotlights are the most popular wall lights.

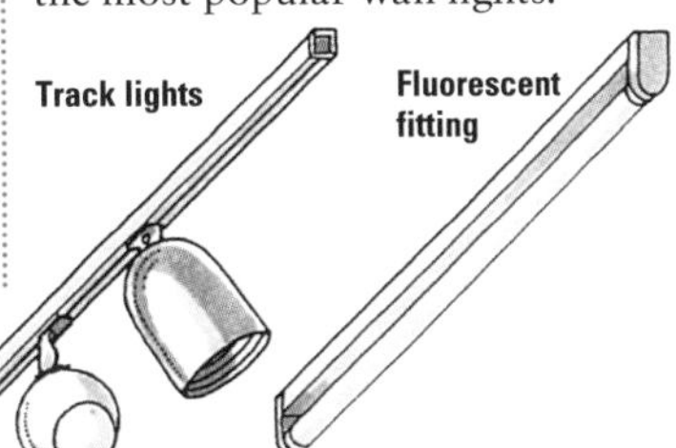

Fluorescent light fittings

A fluorescent light uses a glass tube containing mercury vapour. The voltage makes electrons flow between electrodes at the ends of the tube and bombard an internal coating – which fluoresces, producing bright light.

Different types of coating make the light appear 'warmer' or 'cooler'. For domestic purposes, choose either 'warm white' or 'daylight'.

The light fitting, which includes a starter mechanism, is usually mounted directly on the ceiling – though, as they produce very little heat, fluorescent lights are also frequently fitted to the underside of cupboards above kitchen work surfaces.

Striplights

These slim lights are often mounted above mirrors and inside cupboards and display cabinets. They can be controlled by separate microswitches so the light comes on each time the cupboard door is opened. Striplights usually take 30W or 60W tubular tungsten filament bulbs with a metal cap at each end.

Some under-cupboard strip lights are designed to be linked with short lengths of cable so that they can all be powered from a single 13amp plug.

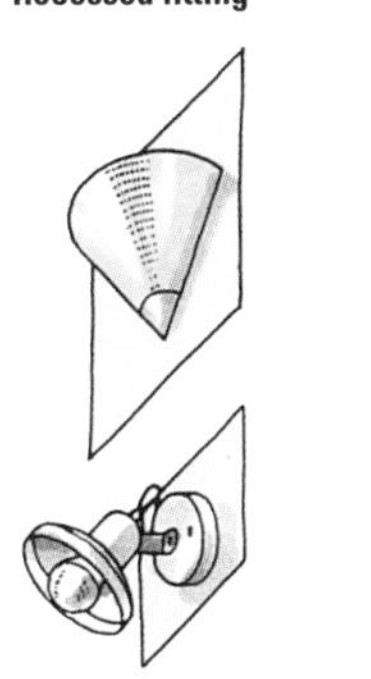

Picture lights
Use discreet light fittings to illuminate individual prints and paintings.

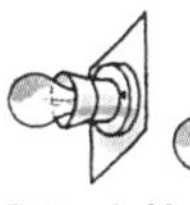

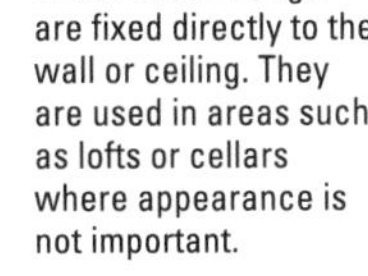

Batten holders
These basic fittings are fixed directly to the wall or ceiling. They are used in areas such as lofts or cellars where appearance is not important.

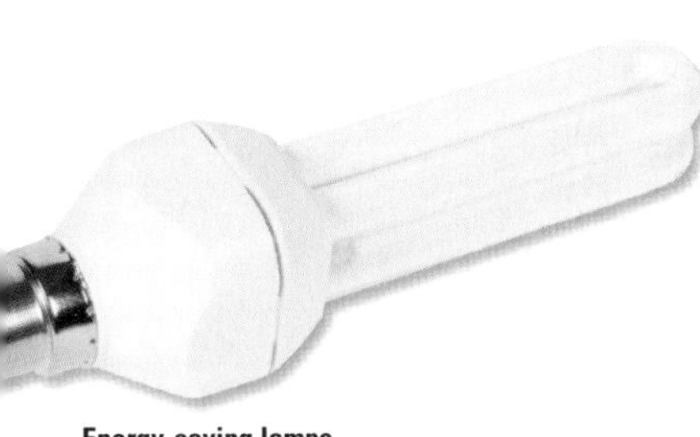

Energy-saving lamps
Compact fluorescent lamps can be used in place of conventional bulbs. They last up to eight times longer and use up to 80 per cent less electricity. The tube is folded to make a very small unit. The lamps either have built-in control circuits or are supplied with plug-in holders containing these controls.

A bewildering choice
It can be difficult to choose a light fitting that meets your needs exactly, but being aware of the main categories helps eliminate fittings that are totally unsuitable.

SEE ALSO: Replacing lampholders 15, Switching off 16, Close-mounted lights 50, Track lighting 50, Fluorescent lights 51, Wall lights 55

Close-mounted lights

Close-mounted light fittings often have a backplate that screws directly to the ceiling, in place of a ceiling rose. To fit one, switch off the power at the consumer unit and take out the circuit fuse or lock off the MCB, then remove the ceiling rose and fix the backplate to the ceiling.

If only one cable feeds the light, attach its conductors to the terminals of the lampholder and connect the earth wire to the terminal on the backplate.

Since more heat is generated inside an enclosed fitting, slip heat-resistant sleeving over the conductors before you attach them to their terminals.

If the original ceiling rose was wired into a loop-in system, then you will find that a close-mounted light fitting won't accommodate all the cables. In which case, withdraw the cables into the ceiling void and wire them into a junction box screwed to a length of 100 x 25mm (4 x 1in) timber nailed between the joists; then run a short length of heat-resistant cable from the junction box to the new light fitting.

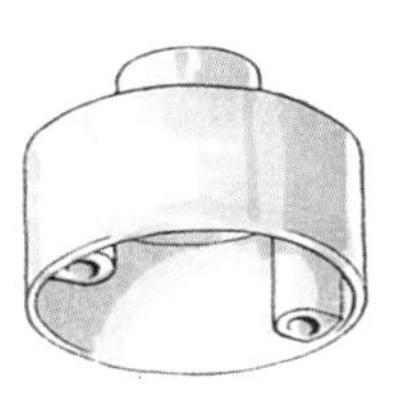

1 BESA box
Use a BESA box (also known as a conduit box) to house the connections when a light fitting is supplied without a backplate. A metal box must be earthed.

FITTING A PLASTIC BESA BOX

Fix a wooden platform between the ceiling joists to support the junction box and the plastic BESA box.

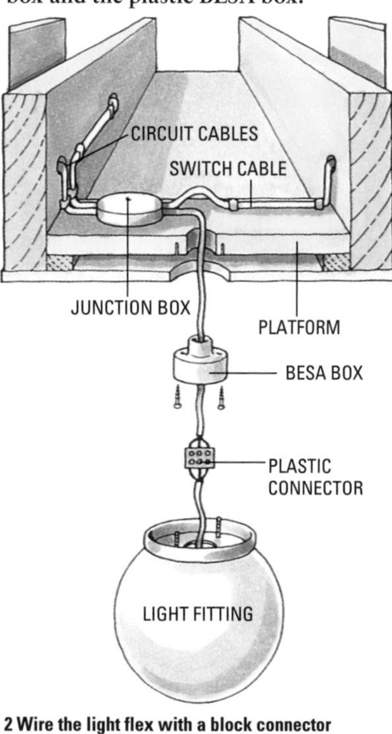

2 Wire the light flex with a block connector

Fittings without backplates

Sometimes close-mounted lights are supplied without backplates.

Wiring Regulations stipulate that all unsheathed wires and terminals have to be enclosed in a noncombustible housing – so if you plan to use a fitting without a backplate, you must find a means of complying. The best way is to fit a BESA box **(1)**, a plastic or metal box that is fixed into the ceiling void so as to lie flush with the ceiling (see below left).

The screw-fixing lugs on the box should line up with the fixing holes in the light fitting's coverplate, but check that they do so before buying the box. You will also need two machine screws of the appropriate thread for attaching the light to the BESA box.

Check that there isn't a joist directly above where you wish to fit the light (if there is one, move the light to one side until it fits between two joists). Then hold the box against the ceiling, trace round it, and carefully cut the traced shape out of the ceiling with a padsaw.

Cut a platform from timber 25mm (1in) thick to fit between the joists, and place it directly over the hole in the ceiling while an assistant marks out the position of the hole on the board from below. Then drill a cable-feed hole centrally through the shape of the ceiling aperture marked out on the board. If there's a boss on the back of the BESA box, the hole must be able to accommodate it. Position the box and screw it securely to the platform.

Have your assistant press some kind of flat panel against the ceiling and over the aperture. Fit the BESA box into the aperture from above, so that it rests on the panel; drop the platform over the BESA box, and mark both ends on both joists. Screw a batten to each joist to support the platform at that level. Fix the platform to the battens and feed the cable through the hole in the centre of the BESA box.

For attaching the cable conductors, the light fitting will probably have a plastic connector **(2)**, which may have three terminals. Alternatively, there may be a separate terminal for the earth conductor attached to the coverplate. After securing the conductors, fix the coverplate to the BESA box with the machine screws.

If the original ceiling rose was fed by more than one cable, connect them to a junction box in the ceiling void, as described above left.

Fitting a downlighter

Decide where you want the light, check from above that it falls between joists, and then use the cardboard template supplied with all downlighters to mark the outline of the circular aperture in the ceiling. Drill a series of 12mm (½in) holes just inside the perimeter of the marked circle to remove most of the waste, then cut it out with a padsaw.

Bring a single lighting-circuit cable from a junction box through the opening and attach the cable to the downlighter, following the maker's instructions. You may have to fit another junction box into the void in order to connect the circuit cable to the heat-resistant flex attached to the light fitting.

Fit the downlighter into the opening and secure it there by adjusting the clamps that bear on the hidden upper surface of the ceiling.

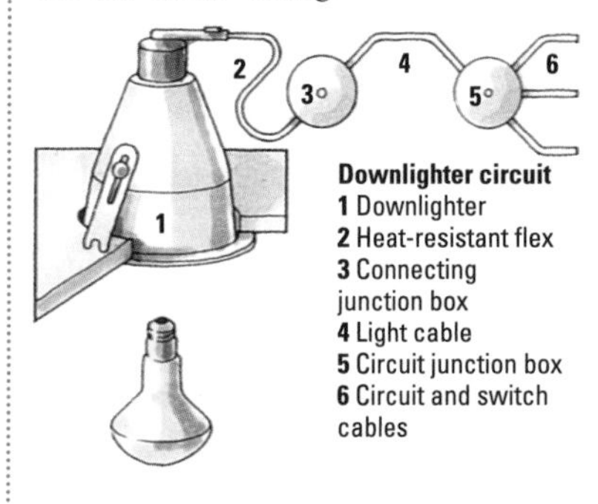

Downlighter circuit
1 Downlighter
2 Heat-resistant flex
3 Connecting junction box
4 Light cable
5 Circuit junction box
6 Circuit and switch cables

Fitting track lighting

Ceiling fixings are supplied with all track-lighting systems. Mount the track so that the terminal-block housing at one end is situated close to where the old ceiling rose was fitted. Pass the circuit cable into the fitting and wire it to the cable connector provided. If the circuit is a loop-in system, mount a junction box in the ceiling void to connect the cables.

Make sure that the number of lights you intend to use on the track will not overload the lighting circuit – which can supply a maximum of eleven 100W lamps or their equivalent.

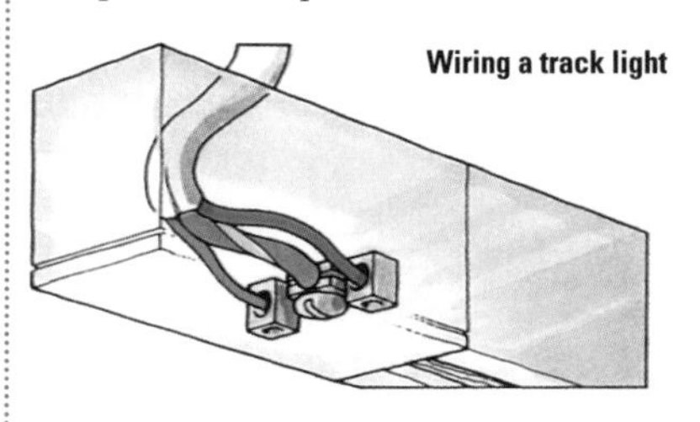

Wiring a track light

SEE ALSO: Switching off 16, Loop-in system 47, 48, Fixing to ceiling 48, Junction box 48, Close-mounted lights 49, Recessed lights 49, Track lights 49

Light switches

Remove the ceiling rose and then screw the new light fitting to the ceiling, positioned so that the circuit cable can be fed into it conveniently.

Fluorescent light fittings are supplied with terminal blocks for connection to the mains supply. Each terminal block will take only three conductors – so either the fitting must be connected to a junction-box system or a junction box must be installed in the ceiling void to accommodate loop-in wiring, as for a close-mounted light (see opposite). Fluorescent lights normally need earth connections, so they can't be used with old systems that lack earth conductors.

You can mount a fluorescent unit by screwing directly into the ceiling joists or into boards nailed between the joists to provide secure fixings.

Wiring a fluorescent light fitting
A simple plastic block connector for the circuit cable is fitted inside a fluorescent light fitting.

Fluorescents under cupboards (N)

You can fit fluorescent lighting underneath wall-mounted kitchen cupboards to illuminate the work surfaces below. The power is supplied from a switched fused connection unit fitted with a 3amp cartridge fuse.

When installing a second fluorescent light fitting, you can supply its power by wiring it into the terminal block of the first one.

The type of switch that's most commonly used for lighting is the plate switch. This has a switch mechanism mounted behind a square faceplate with either one, two or three rockers. Although that's usually enough for domestic purposes, double faceplates with as many as four or six rockers are also available.

A one-way switch simply turns a light on and off, but two-way switches are wired in pairs so that the light can be controlled from two places – typically, at the head and foot of a staircase. It's also possible to have an intermediate switch, to allow a light to be controlled from three places.

Any type of switch can be flush-mounted in a metal box buried in the wall or surface-mounted in a plastic box. Boxes 16 and 25mm (⅝ and 1in) deep are available, to accommodate switches of different depths.

Where there is not enough room for a standard switch, a narrow architrave switch can be used. There are double versions with two rockers, one above the other.

As well as turning the light on and off, a dimmer switch controls the intensity of illumination. Some types have a single knob that serves as both switch and dimmer. Others incorporate a separate knob for switching, so the light level does not have to be adjusted every time the light is switched on.

The Wiring Regulations forbid the positioning of a conventional switch within reach of a washbasin, bath or shower unit – so only ceiling-mounted double-pole switches with pull-cords must be used in bathrooms.

Fixing switches and running cable
Light switches need to be installed in relatively accessible positions, which normally means just inside the door of a room, at about adult shoulder height.

In order to reach the switch, lighting cable is either run within hollow cavity walls or buried in the wall plaster.

Methods for fixing mounting boxes in place are similar to those described for fitting socket outlets.

Choosing switches

Most light switches are made from white plastic, but you can buy other finishes to compliment your decorative scheme. Bright primary-coloured switches can look striking in a modern house, while reproductions of antique brass switches are both appropriate and attractive in a traditional interior.

● **Double-pole switches**
With this type of switch, both live and neutral contacts are broken when it is off.

Selection of light switches
1 One-gang rocker switch
2 Two-gang rocker switch
3 Primary-coloured rocker switch
4 Reproduction antique switch
5 One-gang dimmer switch
6 Two-gang dimmer switch
7 Touch dimmer switch
8 Two-gang architrave switch
9 Ceiling switch

☛ SEE ALSO: Building Regulations 8, Bathroom zones 11, Switching off 16, Running cable 23–5, Mounting boxes 28–9, Fused connection units 34, Fixing to ceiling 48, Junction box 48, Fluorescents 49, Wiring switches 52, 54

Replacing switches

Replacing a damaged switch is a matter of connecting the existing wiring to the terminals of the new switch – making sure that you connect the wires in exactly the same way as in the old one.

Check that a new faceplate for a surface-mounted switch will fit the existing mounting box; otherwise, you will have to replace both parts of the switch. If you are able to use the box, attach the new faceplate with the old machine screws. You can then be certain of having screws that will match the threads.

If you want to replace a surface-mounted switch with a flush-mounted one, remove the old switch, then hold the metal box over the position of the original switch and trace round it. Cut away the plaster to the depth of the box, then screw it to the brickwork. Take great care not to damage the existing wiring while you are working.

• Switching off
Always turn off the power and remove the relevant fuse or remove (or lock off) the MCB before you take off a switch faceplate to inspect the wiring.

• Old colour coding
When working on a house built before 2005 you are likely to find that the existing cables are colour-coded black for neutral and red for live. The diagrams on this page show old-style switch cables.

Replacing a one-way switch

Examine a one-way switch and you will see that it is serviced by a two-core-and-earth cable. The earth conductor, if there is one, will be connected to an earth terminal on the mounting box. The red and black conductors will be connected to the switch itself.

A true one-way switch has only two terminals, one situated above the other, and the red or black conductors can be connected to either terminal **(1)**. The back of the faceplate is marked 'top' to ensure that you mount the switch the right way up, so the rocker is depressed when the light is on. The switch would work just as well upside down – but the 'up for off' convention is a useful one, as it tells you whether the switch is on or off even when the bulb has failed.

You may come across a light switch that is fed by a two-core-and-earth cable and operates as a one-way switch yet has three terminals **(2)**. This is a two-way switch that has been wired for one-way function – something that's fairly common and perfectly safe. If the switch is mounted the right way up, then the red and black wires should be connected to the 'Common' and 'L2' terminals **(2)** – either wire to either terminal.

Replacing a two-way switch

A two-way switch will have at least one conductor in each of its three terminals.

Without going into the complexities of two-way wiring at this stage, you will find that the most straightforward method of replacing a damaged two-way switch is simply to make a written note of which conductors run to which terminals before you start to disconnect the various wires.

Another simple method is to detach the wires from their terminals one at a time, and connect each one to the corresponding terminal on the new two-way switch before you deal with the next conductor.

Two-gang switches

A two-gang switch is the name for two individual switches mounted on a single faceplate. Each of the switches may be wired differently. One may be working as a one-way switch, and the other as a two-way **(3)**. To transfer the wires from an old switch to the terminals of a new one, work on one switch at a time and use one of the methods for replacing a two-way switch described above.

Replacing a rocker switch with a dimmer switch

Examine the present switch in order to determine the type of wiring that feeds it, then purchase a dimmer switch that will accommodate the existing wiring. The manufacturers of dimmer switches provide instructions with them, but the connections are basically the same as for ordinary rocker switches **(4)**.

Don't use a dimmer switch to control a fluorescent light.

HOW SWITCHES ARE WIRED

It is very easy to replace a damaged switch or to swap one for a different type of switch. The illustrations below show four common methods of wiring switches. If a switch appears to be wired differently, it is probably part of a two-way or three-way lighting system. Replace switches as described left.

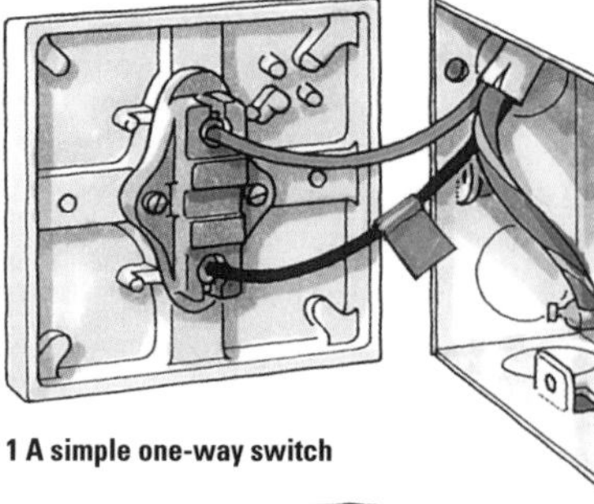

1 A simple one-way switch

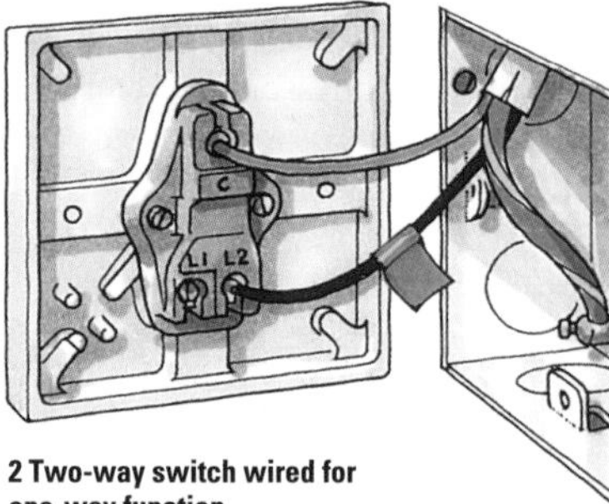

2 Two-way switch wired for one-way function

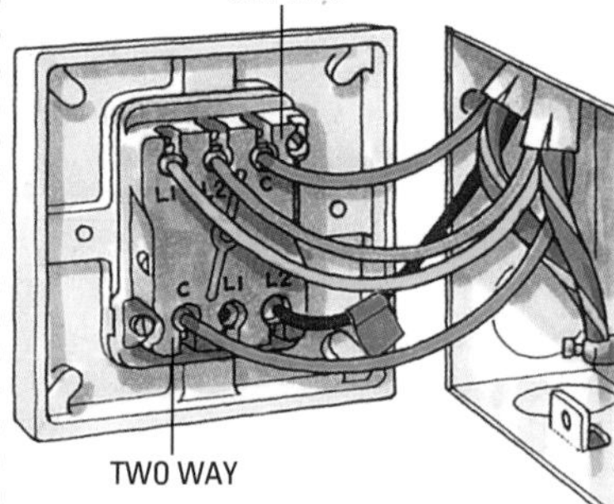

3 Two-gang switch for one-way and two-way functions

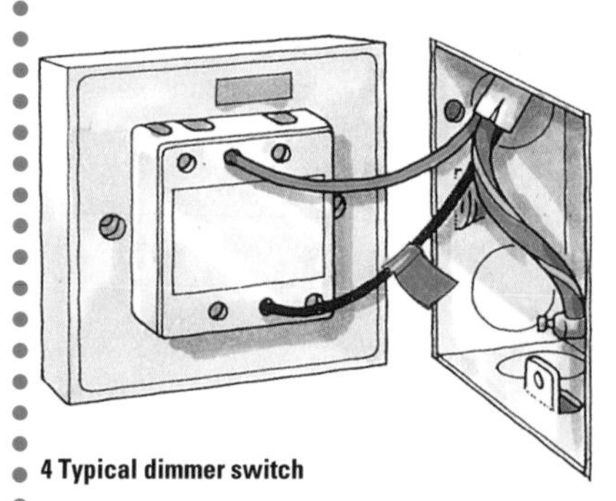

4 Typical dimmer switch

SEE ALSO: Switching off 16, Cables 22, Flush mounting 29, Switches 51, Two/three-way lighting 54

Adding new switches and circuits

When you want to move a switch or install a new one, you will have to modify the circuit cables or run a new spur cable from the existing lighting circuit to take the power to where it is needed.

Replacing a wall switch with a ceiling switch Ⓝ

In a bathroom, light switches must be outside zones 0 to 3. If your bathroom has a wall switch that breaks this rule, replace it with a ceiling switch that is at least 0.6m (2ft) horizontally from the bath or shower.

Turn the power off at the consumer unit, then remove the old switch. If the cable running up the wall is surface-mounted or in a plastic conduit, you can pull it up into the ceiling void. It should be long enough to reach the point where the new switch is to be located.

If the switch cable is buried in the wall, trace it in the ceiling void and cut it. Then wire the remaining part that runs to the light into a three-terminal junction box fixed to a joist or to a piece of wood nailed between two joists. Connect the conductors to separate terminals **(1)**, and from those terminals run a new 1mm^2 two-core-and-earth cable to the site of the ceiling switch.

Bore a hole in the ceiling to pass the cable through to the switch. Screw the switch to a joist if the hole is close enough; otherwise, fix a support board between the joists.

Knock out the entry hole in the back-plate of the switch and pass the cable through it, then screw the plate to the ceiling. Strip and prepare the ends of the conductors, connecting the earth to the terminal on the backplate. Connect the brown and blue conductors to the terminals on the switch – either wire to either terminal **(2)**. Finally, attach the switch to the backplate and make good any damage done to the plasterwork.

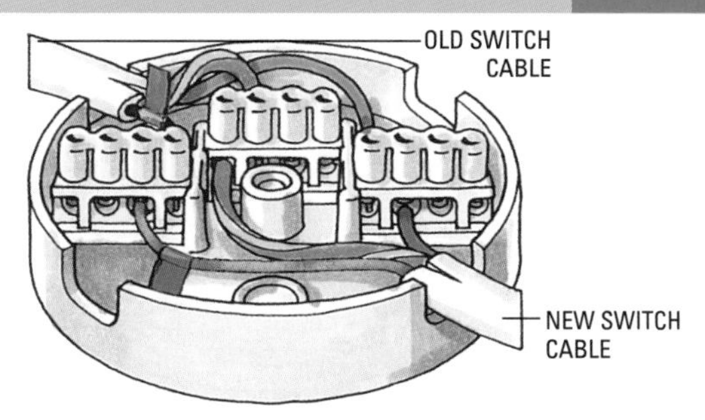

1 Link the switch cable with a junction box

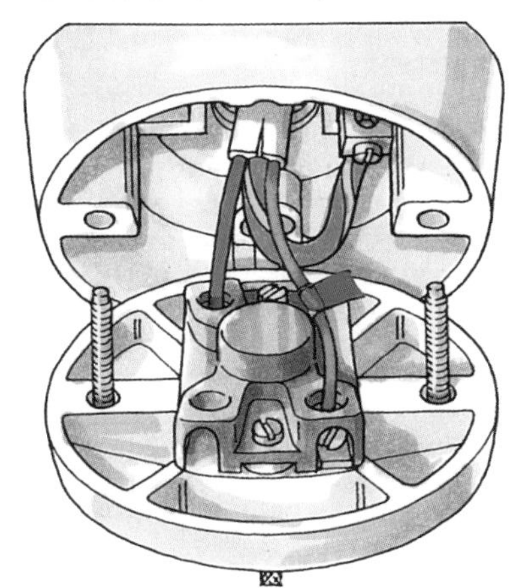
2 Wiring a ceiling switch

• Old colour coding
When working on a house built before 2005 you are likely to find that the existing cables are colour-coded black for neutral and red for live. The diagrams on this page show new-style cables being connected to existing old-style circuit cables.

Adding a new switch and light

Switch the power off at the consumer unit and inspect your lighting circuit to check whether it is earthed. If there's no earth wire, get expert advice before installing new light fittings.

Decide where you want to mount the light, and bore a hole through the ceiling for the cable. If the ceiling rose can't be screwed to a nearby joist, nail a board between two joists to provide a strong fixing for the rose.

Bore another hole in the ceiling right above the site of the new switch and as close to the wall as possible. Push twists of paper through both holes, so you can find them easily from above.

Screw the switch mounting box to the wall and cut a chase in the plaster for the cable, up to the appropriate hole already bored in the ceiling.

Your new light fitting can be supplied with power from a nearby junction box or ceiling rose that's already on the lighting circuit – or, if it's more convenient, from a new junction box wired into the lighting-circuit cable.

From whichever of these sources you choose, run a length of 1mm^2 two-core-and-earth cable to the position of the new light fitting – but don't connect to the lighting circuit till the whole of the new installation has been completed. Push the end of the cable through the hole in the ceiling and identify it with tape marked 'Mains' **(1)**.

The next step is to run a similar cable from the switch to the same lighting point **(1)**.

Strip and prepare the cable at the switch – connecting the earth wire to the terminal on the mounting box – and connect the brown and blue conductors, either wire to either terminal if it is a one-way switch. If you are using a two-way switch, connect the brown wire to the 'Common' terminal and the blue wire to 'L2'. Now screw the switch to the mounting box.

Knock out the cable-entry hole in the ceiling rose and feed both cables through it, then screw the rose to the ceiling.

Take the cable marked 'Mains' and connect its brown conductor to the live central block and its blue one to the neutral block. Slip a green-and-yellow sleeve over the earth wire and connect it to the earth terminal.

Connect the brown conductor of the switch cable to the live block, and the blue wire to the switch-wire block: mark the blue wire with brown tape. Connect the switch earth wire to the common earth terminal. Connect the pendant flex and screw on the rose cover.

Make sure the power is turned off, then connect the new light circuit to the old one at the rose or junction box. The new conductors will have to share terminals with the old wires already connected: brown to live, blue to neutral, and earth to earth **(2)**. Finally, test and switch on the new circuit.

1 Identify the mains cable

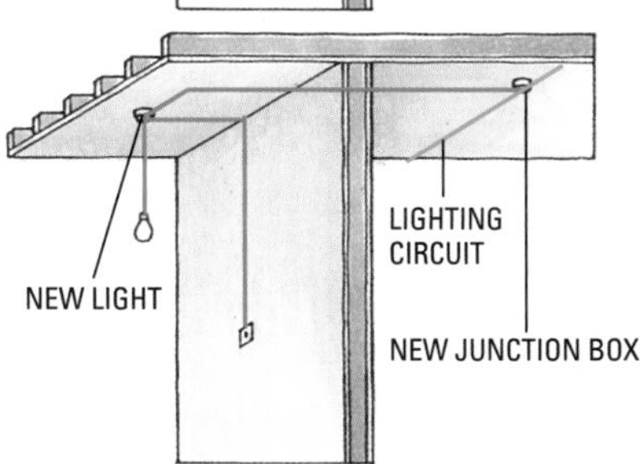

Circuit for a new light ▶
You can take the power for a new light from an existing ceiling rose or junction box, or insert a new junction box into the existing lighting circuit.

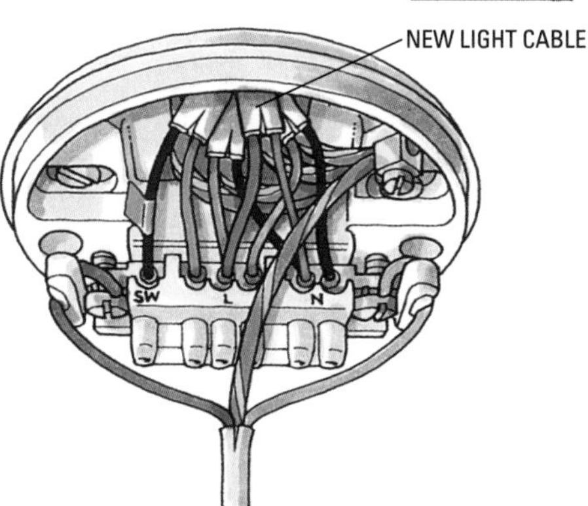

2 Lighting cable connected to a loop-in rose

Two- and three-way lighting

Adding a two-way light

There are several situations in which a light should be controllable from two points. For example, in a long entrance hall the light is best switched from both ends of the passageway; and a landing light needs to be controlled from both the top and bottom of the stairs.

Installing a new two-way light is very similar to installing a one-way light, the only real difference being in the wiring of the switches.

First, mount the ceiling rose and both of the two-way switches, then run $1mm^2$ two-core-and-earth cable from the power source to the light and from the light to the nearest switch. Don't connect the new installation to the lighting circuit until all the wiring has been completed.

Run a $1mm^2$ three-core-and-earth cable from the first to the second switch. Then strip the conductors and prepare them for connecting to the switches, slipping insulating sleeves over the bare earth wires.

At the first switch you will have two cables to connect: the switch cable from the light and the one linking the two switches. The switch cable has three conductors (brown, blue and green-and-yellow); the linking cable has four (brown, black, grey and green-and-yellow). Connect the green-and-yellow wires from both cables to the earth terminal on the mounting box **(1)**. Next, connect the brown wire from the linking cable to the 'Common' terminal on the switch. Connect the black wire and either the brown or blue switch-cable wire to the 'L1' terminal. Connect the grey wire and the remaining switch-cable wire to 'L2' **(1)**. Screw the switch's faceplate to the mounting box.

At the second switch, connect the linking cable's green-and-yellow wire to the earth terminal; its brown wire to the 'Common' terminal; its black wire to 'L1'; and its grey wire to 'L2' **(1)**. Screw the switch's faceplate to the box.

Make sure the power is switched off, and then connect the installation to the lighting circuit at either a ceiling rose or a junction box. Finally, test the new installation.

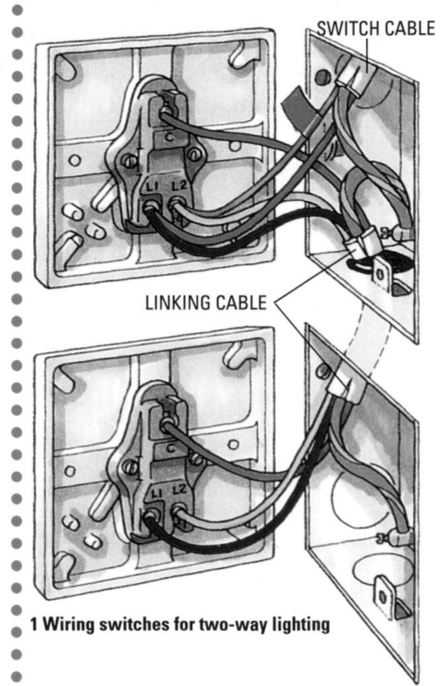

1 Wiring switches for two-way lighting

Three-way lighting

You can control a light from three places by adding an intermediate switch to the circuit described above.

The intermediate switch interrupts the three-core-and-earth cable linking the other two. It has two 'L1' and two 'L2' terminals.

At its mounting box you will have two identical sets of wires – brown, black, grey and green-and-yellow. Connect the green-and-yellow wires to the earth terminal on the box **(2)** and join the two brown wires – which play no part in the intermediate switching – with a plastic block connector **(2)**. Ease the block to one side, in order to clear the switch when you fit it.

Connect the grey and black wires of either cable to the 'L1' terminals on the new switch and those of the other cable to the 'L2' terminals **(2)**. Then screw the faceplate to the mounting box.

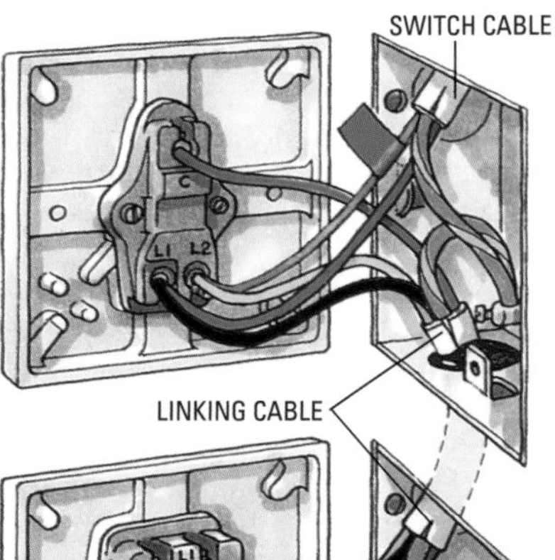

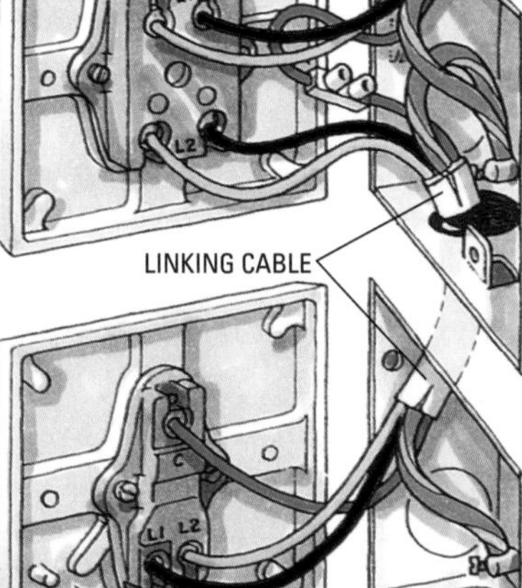

2 Wiring switches for three-way lighting

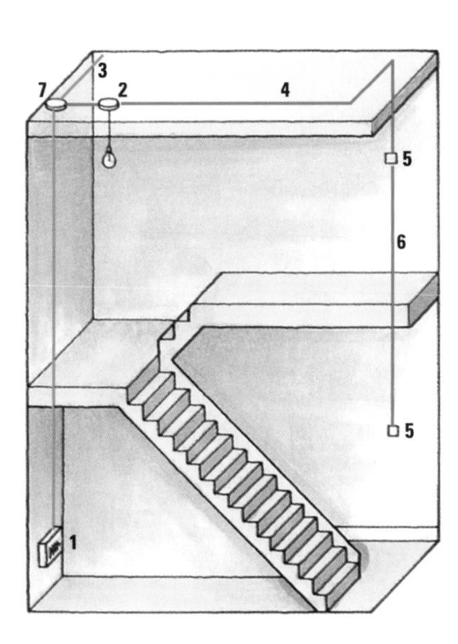

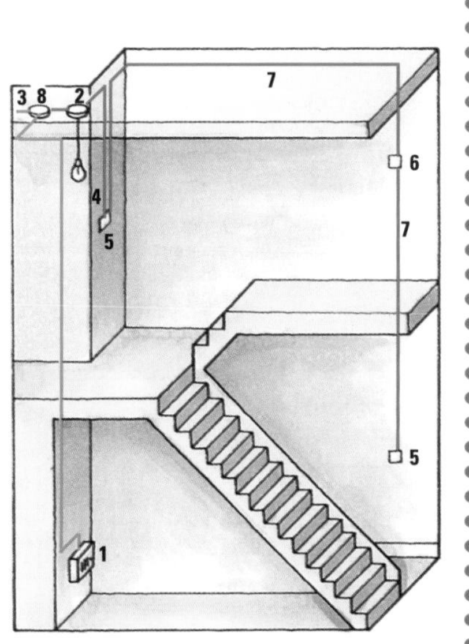

Two-way-lighting circuit *(right)*
1 Consumer unit
2 Light fitting
3 Lighting-circuit cable
4 Switch cable
5 Switch
6 Linking cable
7 Junction box

Three-way-lighting circuit *(far right)*
1 Consumer unit
2 Light fitting
3 Lighting circuit
4 Switch cable
5 Switch
6 Intermediate switch
7 Linking cable
8 Junction box

SEE ALSO: Switching off 16, Connecting to a junction box 31, Lighting circuits 47, Adding a new switch and light 53, Connecting to a loop-in rose 53, Circuit lengths 66

Adding wall lights

Many wall lights are supplied without integral backplates to enclose the wires and connections. To comply with the Wiring Regulations, such a fitting must be attached to a noncombustible mounting such as a BESA box – a round plastic or metal box that's screwed to the wall in a recess chopped out of the plaster and brickwork.

Alternatively, you can use an architrave-switch mounting box. This is a slim box that leaves plenty of room on each side for the wallplug fixings needed for the light fitting. Both types of mounting box are fixed to the wall the same way as flush-mounted sockets.

The basic circuit and connections

The simplest way to connect wall lights to the lighting circuit is via a junction box. The procedure is to complete the wall-light installation first, then switch off the electricity and connect the new installation with the junction box.

Wire up a one-way switch. All the wall lights in the room will be controlled by this switch – although if you choose lights with integral switches they can be controlled individually, too.

Next, run a 1mm^2 two-core-and-earth cable from the junction box, looping in and out of each wall-light mounting to the last one, where the cable ends.

Prepare the cut ends of the conductors for connection. At each of the lights, slip green-and-yellow sleeving over the earth wires and connect them to the earth terminal on the mounting box **(1)**.

Connect up the brown and blue wires to the block connector inside each light fitting – connecting the blue conductors to the terminal already holding the blue wire, and the brown conductors to the terminal already holding the brown wire **(1)**.

The last wall-light mounting will have one end of the cable entering it. Strip and prepare the ends of the wires, then connect them as described above.

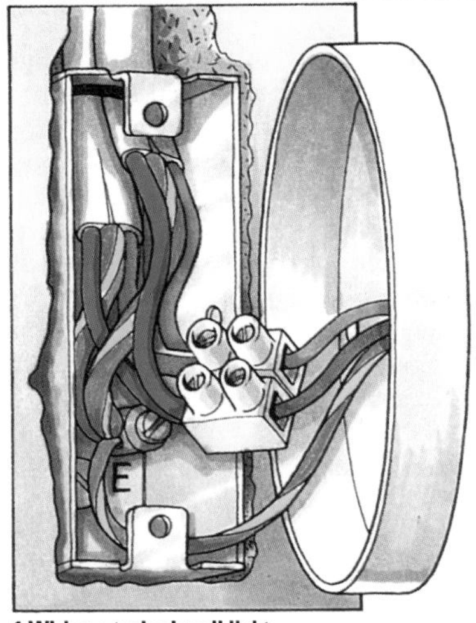

1 Wiring a typical wall light

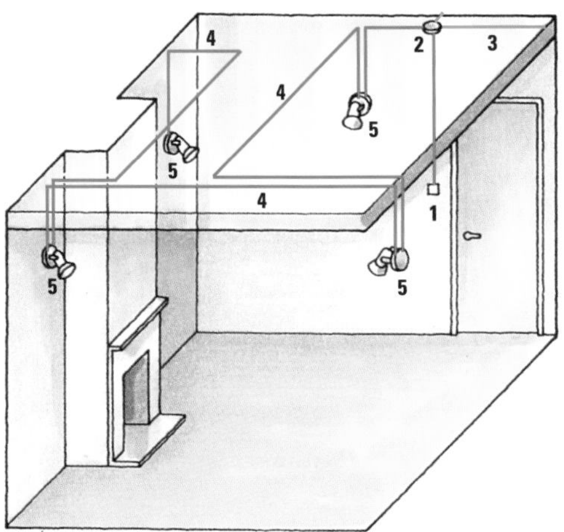

Basic wall-light circuit
The basic circuit and connections are as described above.
1 Switch
2 Junction box
3 Existing lighting circuit
4 Wall-light cable
5 Wall light

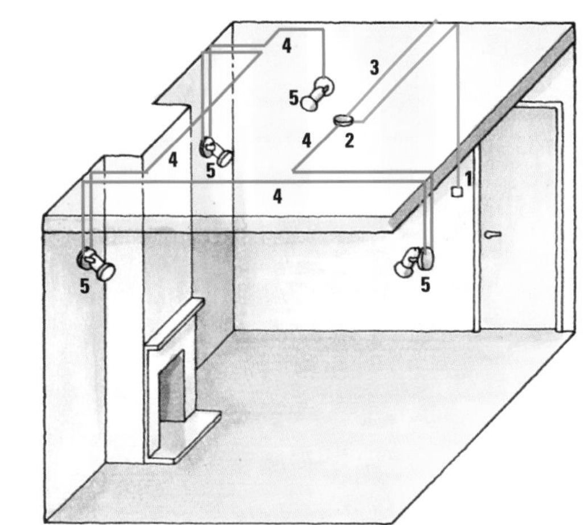

Replacing a ceiling light
You can dispense with a ceiling light in favour of wall lights, using the existing wiring and switch. Switch off the power, then remove the rose and connect up the wiring to a fixed junction box.
1 Existing switch and cable
2 Junction box replaces rose
3 Existing lighting circuit
4 1mm^2 wall-light cable
5 Wall light

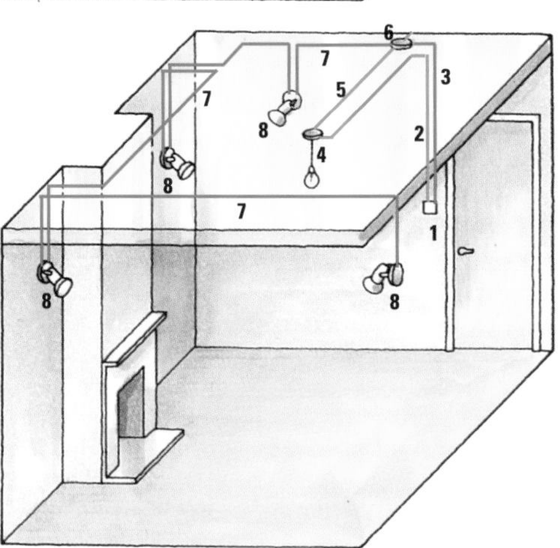

Ceiling light plus wall lights
If you want to retain your ceiling light, you can substitute a two-gang switch for the single one – and wire the present ceiling-light cable to one half of the switch, and the new wall-lighting cable to the other half.
1 Two-gang switch
2 Old switch cable
3 New switch cable
4 Ceiling light
5 Existing lighting circuit
6 Junction box
7 1mm^2 wall-light cable
8 Wall light

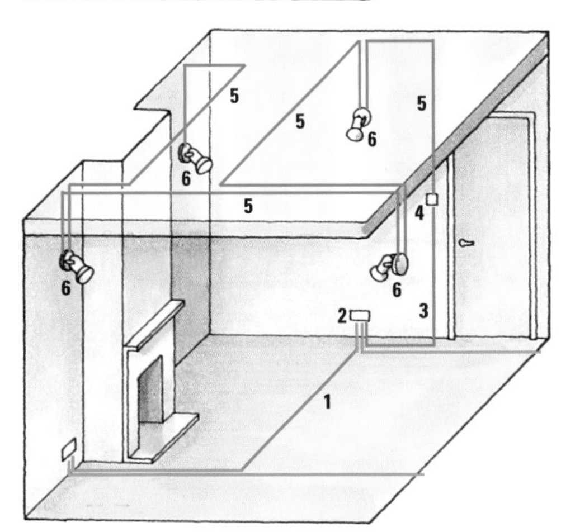

Using a spur
Wall lights can be wired to a ring circuit by means of a spur cable. Run a 2.5mm^2 two-core-and-earth spur from a nearby socket to a switched fused connection unit that has a 3amp fuse.
1 Ring circuit
2 Socket outlet
3 Spur cable
4 Fused connection unit
5 1mm^2 wall-light cable
6 Wall light

SEE ALSO: Switching off 16, Running cable 23–5, Flush mounting 29, Running a spur 31, Fused connection unit 34, Lighting 47, Junction box 48, BESA box 50, One-way switch 52, Two-gang switch 52, Circuit lengths 66

Low-voltage lighting

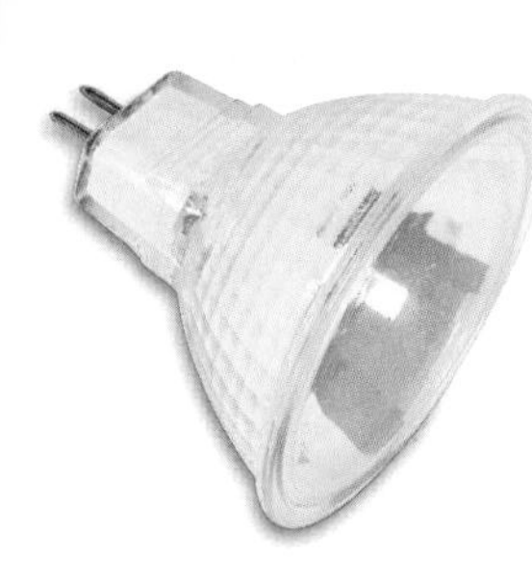

Low-voltage halogen bulbs
The specially designed miniature bulb is the key to what makes low-voltage lighting so attractive. The light source is concentrated into a small filament, which enables accurate focusing of spotlight beams. The integral 'dichroic' reflector allows the heat generated by the filament to escape backward into the fitting, creating a cool but intense white light. Coloured bulbs are also available, for special effects and mood lighting.

Originally, low-voltage halogen light fittings were developed for illuminating commercial premises. Being small and unobtrusive, they blend into any scheme, and the bright intense beams of light that they produce are ideal for display lighting. The potential for dramatic effects and narrowly focused task lighting was not lost on home owners, and manufacturers were quick to respond with a range of low-voltage fittings.

Although installing low-voltage lighting using the methods shown opposite is classified in the Building Regulations as notifiable work, this is not the case when fitting complete ready-connected kits.

Low-voltage light fittings

Perhaps the most widely used low-voltage light fittings are miniature fixed or adjustable 'eyeball' downlighters, recessed into the ceiling. They can be mounted individually or wired in groups to a transformer, which is also concealed in the space above the ceiling. Some fittings are made with integral transformers; these include table lamps and small spotlights that can be mounted on the wall or ceiling. Others, such as track lights, combine several individual fittings connected to a single transformer. Unique to low-voltage lighting are fittings connected to exposed plastic-sheathed cables suspended across the ceiling or wall.

Using the chart
Decide how many 20W or 50W bulbs you are going to use (column 1). Their combined wattage (column 2) will help you to determine the appropriate transformer (column 3).

Similarly, decide how many bulbs will be supplied by a single cable (some fittings combine more than one bulb). From that number (in column 1), trace across the chart to find which size of cable is safe for the maximum length you require.

				MAXIMUM CABLE LENGTHS (metres)			
		WATTS	Transformer	1.5mm²	2.5mm²	4mm²	6mm²
Number of 20W bulbs	1	20	20-50W	9.9	16.0	26.2	39.5
	2	40	50W	5.0	8.0	13.1	19.7
	3	60	50 or 60W	3.3	5.3	8.7	13.2
	4	80	105W	2.5	4.0	6.5	9.9
	5	100	105W	2.0	3.2	5.2	7.9
	6	120	150W	1.7	2.7	4.4	6.6
	7	140	150W	1.4	2.3	3.7	5.6
Number of 50W bulbs	1	50	50 or 60W	4.0	6.4	10.5	15.8
	2	100	105W	2.0	3.2	5.2	7.9
	3	150	150 or 200W	1.3	2.1	3.5	5.3
	4	200	200 or 250W	1.0	1.6	2.6	3.9
	5	250	250 or 300W	0.8	1.3	2.1	3.2
	6	300	300	0.7	1.1	1.7	2.6

PLANNING THE CIRCUIT

For low-voltage lighting, you need to connect a cable to the 230V mains supply (see opposite) and run this cable to a junction box from which you can run another cable to a switch. From the same junction box, take a cable to the transformer; and from there run separate cables to each light fitting. The circuit is simple, but it needs to be designed carefully in order to optimize the life of the fittings.

Optimum voltage

Even a small increase in the designed voltage can halve the life of a bulb. If the voltage is too low, light output drops and eventually the bulb blackens. Voltage can be affected in a number of ways, and you need to select your equipment accordingly.

Choose a transformer with an output that closely matches the combined wattage of the bulbs on the circuit. It's important to ensure that the total wattage of these bulbs is greater than 70 per cent of the transformer rating, or the bulbs will burn out relatively quickly. For example, a 50W transformer can supply two 20W bulbs or one 50W. A 200W transformer is perfect for four 50W bulbs, but not for six 20W bulbs (for these, you would want a 150W transformer). If you buy a low-voltage kit, you can be sure the transformer is suitable; otherwise, use the chart on the left to help you choose a transformer that meets your needs.

Even with a perfectly matched transformer, replace a blown bulb as soon as possible to avoid overloading the other bulbs on the circuit.

Using ordinary dimmer switches is not advisable, because they too reduce voltage to an unacceptable level. For this type of control, check that the low-voltage fittings are suitable for dimming and use only special dimmer switches designed for low-voltage lighting.

Cable length and size

The length of the cable running from transformer to light fitting is another factor to consider. If the cable is too long, the resulting voltage drop could have a detrimental effect. To ensure the bulbs have equal volt drop, install separate cables running to each light fitting and try to make these cables similar in length.

Low-voltage lamps draw a relatively high current compared with mains-voltage ones, so make sure the cable supplying them is large enough to avoid overheating and prevent an unacceptable drop in voltage. Use the chart on the left to help you select the correct size of cable.

SEE ALSO: Cables 22, Dimmer switches 51

Wiring the circuit

The circuit described here includes a separate transformer, supplying two individual lamps of equal wattage.

Making the connections

Having decided on the position of each lamp, place the transformer midway between the two fittings. Screw the transformer to a ceiling joist, halfway between the ceiling and any floorboards above. Clear any insulation from around the transformer.

Plan the best route to a ceiling rose where you can pick up the 230V mains supply **(1)**. If that is not possible, insert a junction box in the lighting circuit and take the power from there.

Screw a four-terminal junction box to a joist close to the transformer. From the box, run 1mm² two-core-and-earth cable to the wall switch and to the ceiling rose (but don't make these connections yet). Make the connections at the junction box as shown **(2)**.

Run the same size cable from the junction box to the transformer – but before you make the connections, cut off both ends of the earth wire flush with the outer sheathing. Connect the live and neutral conductors to the junction box **(2)**. Connect the other end of the cable to the input (230V) terminals of the transformer **(3)**.

Prepare two cables of suitable size for the light fittings, cropping off the earth wires as before. Run both cables to their respective fittings, keeping them separate from any cable carrying mains-voltage electricity. At the transformer, connect each cable to one of the output terminals as described in the manufacturer's instructions **(4)**.

Each light fitting will have a terminal block attached to the back of the lamp by a short length of heat-resistant flex. Connect the brown and blue wires of the cable to the block **(5)** – either wire to either terminal – then clip the cable to a joist to make sure it cannot touch the back of the light fitting, which can become quite hot.

Fit an ordinary wall switch and make the connections **(6)**.

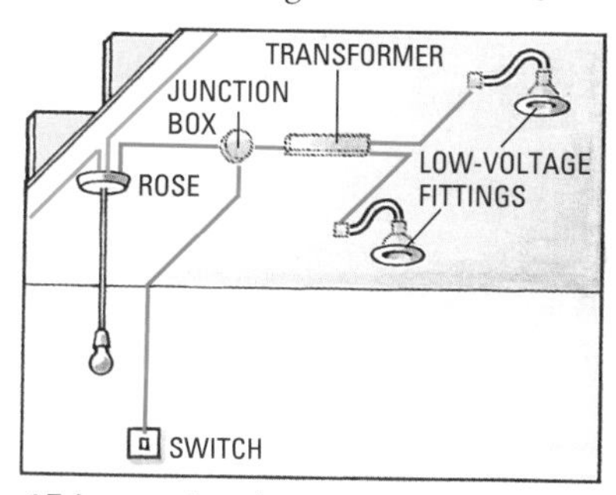

1 Take power from the nearest ceiling rose

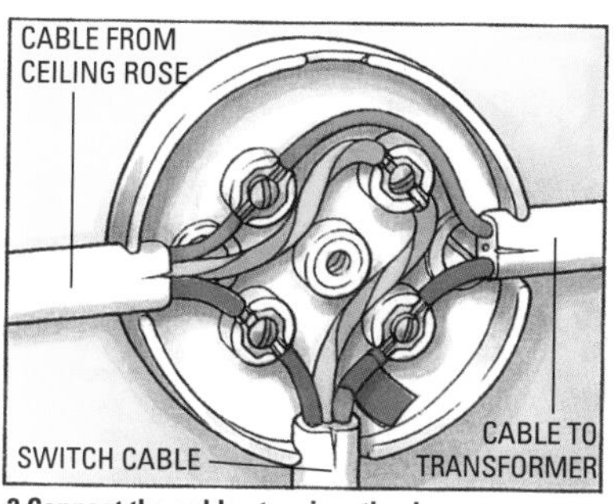

2 Connect the cables to a junction box

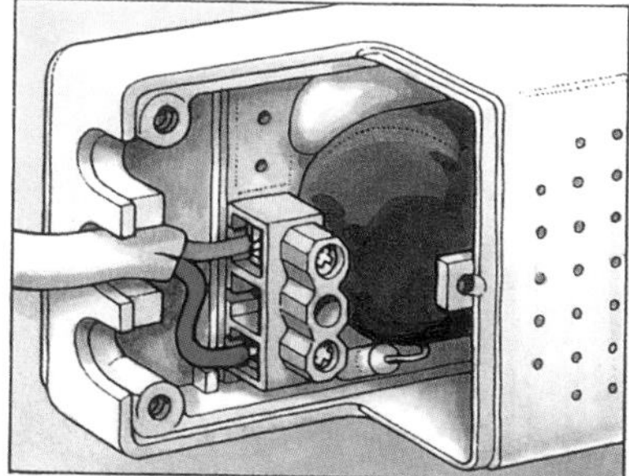

3 Connect the conductors to the input terminals

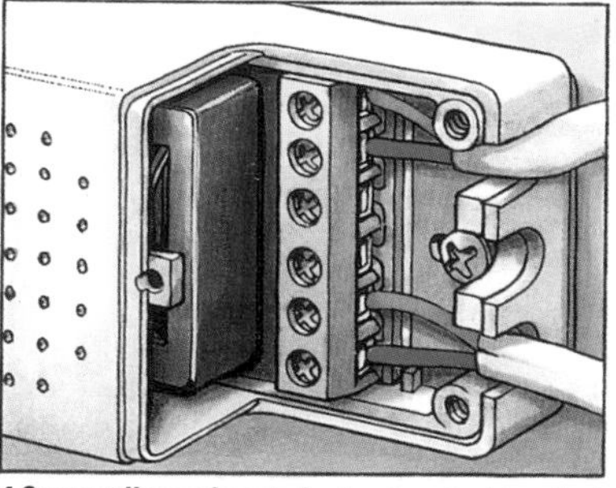

4 Connect live and neutral wires to output terminals

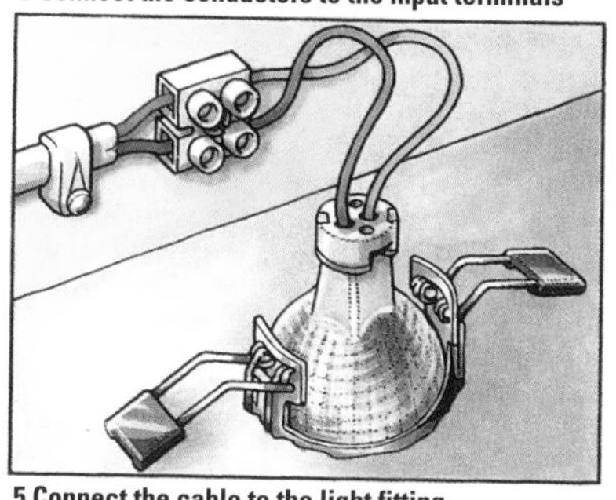

5 Connect the cable to the light fitting

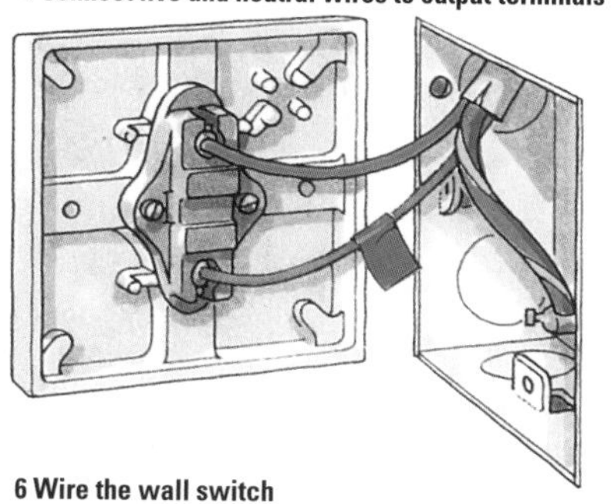

6 Wire the wall switch

MAIN-VOLTAGE HALOGEN

You can buy mains-voltage halogen fittings that are very similar in appearance to the low-voltage versions. Their main advantage is that they don't require a remote transformer to reduce the power to 12V and can therefore be connected to existing wiring, like other mains-voltage fittings. However, the bulbs and fittings are generally more expensive than the low-voltage equivalents.

• 230V halogen
All extra-low-voltage wiring is notifiable, but wiring 230V fittings is only notifiable in special locations.

With some mains-voltage ranges, an electronic transformer is built into the base of each bulb. As a result, the filament within the bulb operates on 12V, just like any other low-voltage fitting. This type of bulb is usually made with an Edison-screw end cap.

Most ranges of mains-voltage fittings accommodate a bulb with two pins that engage a spring-loaded lampholder, similar in principle to the familiar bayonet-cap bulbs. These bulbs contain quartz burners that operate on 230V.

Mains-voltage fittings emit the same sort of bright, sparkling illumination that is normally associated with low-voltage lighting – but if you want a lamp with a relatively cool output, make sure you choose mains-voltage bulbs made with dichroic reflectors, rather than the aluminium-coated versions.

• Old colour coding
When working on a house built before 2005 you are likely to find that the existing cables are colour-coded black for neutral and red for live. The diagrams on this page show new-style cables being connected to existing old-style circuit cables.

Connecting to the mains

Having turned off the power at the consumer unit, connect the live, neutral and earth wires of the circuit-feed cable to the ceiling rose **(7)**.

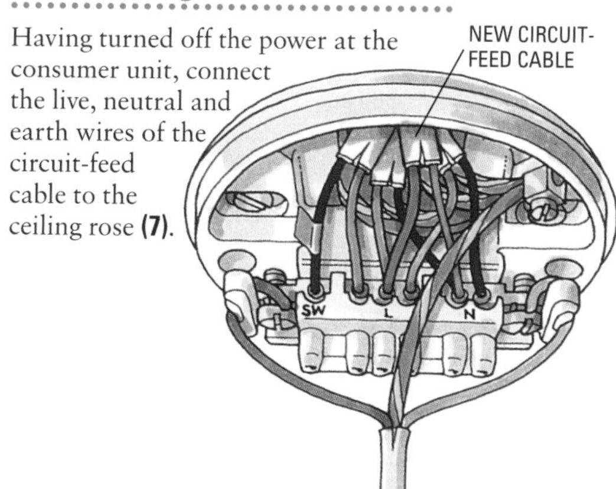

7 Switch off and connect cable to the ceiling rose

SEE ALSO: Building Regulations 8, Switching off 16, Running cable 23–5, Taking power from a junction box 31

Using electricity outdoors

There are good reasons for extending your electrical installation outside the house. First, and most important, it is safer to run electric garden tools from a convenient, properly protected socket than to trail long leads from sockets inside the house – a practice that can lead to serious accidents.

A garage or workshop is also safer, and more efficient, if it is equipped with good lighting and its own circuit from which to run power tools.

Finally, well-arranged spotlights or floodlighting, and waterfalls or fountains powered by electric pumps, add to the charm of a garden or patio and can extend their use in summer by providing a pleasing background for barbecues and outdoor parties.

SAFETY OUTDOORS

The need for absolute safety outdoors cannot be overemphasized. Damp conditions and the fact that users are likely to be in direct contact with the earth can result in fatal accidents if you don't follow the correct procedures.

- Install only light fittings specifically made for outside use.
- Use only cables recommended in the Wiring Regulations, and check their condition regularly.
- Protect all outside installations with residual current devices (RCDs), as they provide an almost instantaneous response to earth-leakage faults.
- Always disconnect the power before servicing electrical equipment and tools. Don't handle pool lighting or pumps unless the power has been switched off.
- Wear thick rubber-soled footwear when using electric garden tools.
- Choose double-insulated power tools for extra protection.

• Cutting through electrical flex
If you accidentally cut the flex that services a power tool, switch off and unplug the tool before you inspect it or touch the severed flex.

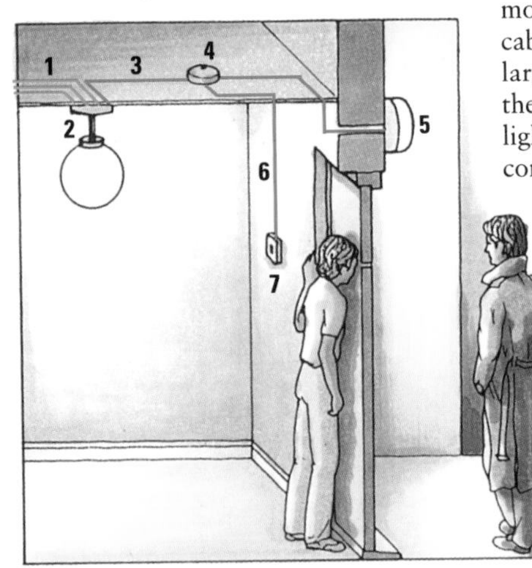

Porch-light circuit
1 Loop-in circuit and switch cables
2 Ceiling rose
3 1mm² lighting cable
4 Junction box
5 Porch light
6 Switch cable
7 Porch-light switch

Fitting a porch light Ⓝ

A light illuminating the front or back entrance welcomes visitors to your home and helps them to identify the house. It also enables you to view unexpected callers before you open the door.

Fit only a light specifically designed for outdoor use. The fitting should be weatherproof, and the lamp or bulb itself should be held in a moisture-proof rubber gasket or cup that surrounds the electrical connections.

If possible, position the porch light in such a way that the cable to it can be run straight through the wall or ceiling of the porch, directly into the back of the fitting. But if you do have to run ordinary cable along an outside wall, it should be protected by being passed through a length of plastic conduit.

Wiring procedure

A porch light is installed by a procedure very similar to that for adding a new light indoors. Take your power from the nearest ceiling rose – probably in the entrance hall – and run it to a 5amp four-terminal junction box screwed to a board between ceiling joists.

From the junction box, run a 1mm² two-core-and-earth cable to a switch mounted near the door, and a similar cable to the light fitting itself. Using a large masonry bit, bore a hole through the wall where you plan to position the light. Cement a short length of plastic conduit into the hole, using a soft rubber grommet to seal each end of the tube. Run the cable through the conduit; wire it into the fitting, following the manufacturer's instructions. Then, with the power switched off, connect the new porch-light cable at the ceiling rose.

INSTALLING A SOCKET FOR GARDEN TOOLS Ⓝ

Many people plug garden tools into the nearest indoor socket – which often results in long extension leads trailing through the house and out into the garden. Any lead that is likely to cause someone to trip is dangerous – and, even more importantly, the Wiring Regulations stipulate that any socket outlet supplying mains power to garden tools or equipment has to be protected by a residual current device (RCD) with a trip rating of 30 milliamps (this also applies to any indoor socket which could reasonably be used to power an outside appliance). The RCD will switch off the power as soon as it detects a fault, long before anyone using the equipment can receive a fatal electric shock.

Outdoors or indoors?

Sockets can be mounted outside, provided they are protected from the weather, although the special procedures involved are best left to a qualified electrician. But you can install a socket in a weatherproof workshop, garage, lobby or conservatory that's part of the house by running a spur from a ring circuit. Mount the socket high enough to prevent it being struck by a wheelbarrow or hidden by garden tools.

Providing RCD protection

You can provide RCD protection in several ways. Perhaps the best method is to have a consumer unit with its own built-in residual current device, or to fit a separate RCD near the consumer unit so that it protects the whole ring circuit, including a spur for garden equipment.

Alternatively, install a socket that incorporates an RCD **(1)**. RCDs fitted in adaptors **(2)** or plugs provide some protection, but they do not satisfy the requirement in the Wiring Regulations for the socket itself to be protected.

1 Socket with built-in RCD
2 Adaptor RCDs plug into any socket outlet

SEE ALSO: Building Regulations 8, Switching off 16, RCDs 17, Running cable 23–5, Running a spur 31, Lighting circuits 47, Junction box 48, One-way switch 52, Adding a new light 53, Connecting to a loop-in rose 53, Circuit lengths 66

Security lighting

Any form of exterior lighting that illuminates the approaches to your house and garage allows you to move about your home with greater convenience and safety – and if it's controlled automatically, it saves you having to fumble with your door keys in the dark. However, probably higher on most people's list of priorities is the added security afforded by installing a system that will detect the presence of intruders and draw attention to their activities.

Dusk-to-dawn lighting

You cannot feel completely secure if you have to remember to switch on exterior lighting every evening. The simplest solution is therefore to install exterior light fittings that are controlled automatically by the ambient light level. Known as dusk-to-dawn lights, these fittings create permanently illuminated areas during the hours of darkness. A photocell detects a change in the level of daylight, switching the lamp on at the approach of darkness and off again early in the morning.

For larger properties, it may be more economical to install a single photocell that controls a number of ordinary exterior light fittings.

The circuit

From a junction box installed in your domestic lighting circuit **(1)**, run a $1mm^2$ two-core-and-earth cable to the light fitting **(2)** and another cable of the same size from the junction box to an ordinary wall switch **(3)**. If you want to install more than one light fitting, run the cable from the junction box to each light in turn, using the single switch to control all of them.

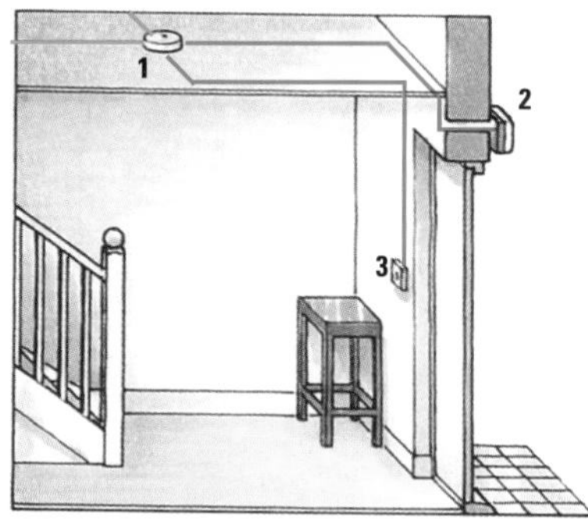

Wiring dusk-to-dawn security lighting
1 Junction box inserted in existing lighting circuit.
2 Light fitting with built-in photocell.
3 Wall switch.

Mounting the light fitting

Follow the manufacturer's fitting instructions, and make sure the light fitting is mounted high enough to prevent unauthorized interference. Drill the cable-access hole through the wall, and line it with a short length of plastic conduit (see opposite). At the same time, bore holes for wallplugs to take the fixing screws provided.

Pass a length of cable through the hole in the wall and into the back of the fitting. Screw the fitting to the wall.

Inside the fitting, cut the separate conductors to length, leaving enough slack to reach their separate terminals. Connect the blue conductor to the neutral terminal and the brown one to the live terminal – these may have internal wiring already connected to them **(1)**. Fit green-and-yellow sleeving over the bare earth conductor and connect it to the earth terminal. If required, connect the internal wires running from the photocell; then fit the bulb in the fitting and replace its cover.

Run the cable from the light fitting along the most convenient route back to where you are going to connect up to the lighting circuit – but do not cut the lighting cable at this stage.

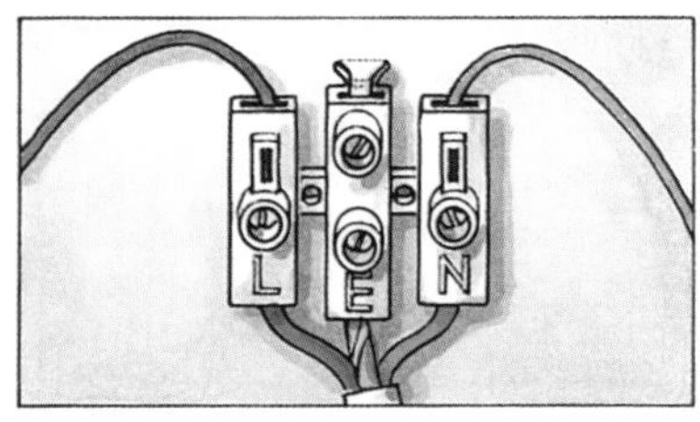

1 Wiring the light fitting

Mounting the light switch

Screw a plastic or metal mounting box to the wall for the switch, and cut a chase in the plaster for the cable. Run the cable into the mounting box, and connect the blue and brown conductors to the terminals of a simple one-way switch **(2)**. Sleeve the earth conductor and connect it to the earth terminal in the mounting box. Screw the faceplate to the mounting box, then take the cable to the point in the lighting circuit where you plan to install the junction box.

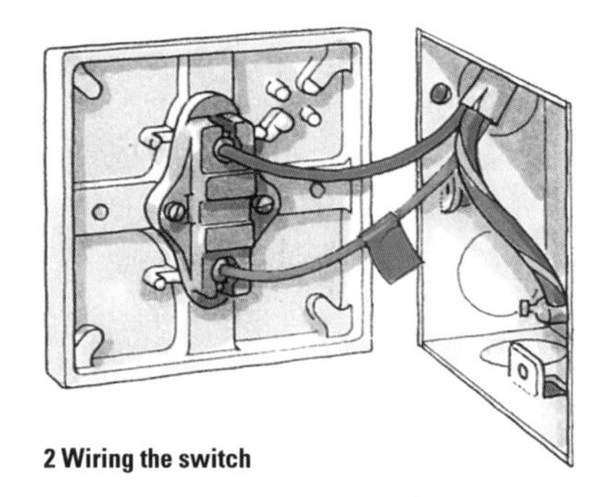

2 Wiring the switch

• Adjusting security lighting
On some dusk-to-dawn light fittings, a screw is provided for adjusting the sensitivity of the photocell. Wait till it is getting dark; then, with the wall switch in the 'on' position, gradually turn the adjustment screw until the light comes on. The photocell will continue to operate your security lighting, provided the wall switch is left on permanently.

Connecting to the lighting circuit

Having first switched off the power at the consumer unit and removed the lighting-circuit fuse or removed (or locked off) the MCB, cut the lighting cable in order to install a four-terminal junction box. Screw the box securely to a joist. The terminals are normally unmarked, so you will need to designate them as live, neutral, earth and switch.

Prepare the cut ends of the lighting-circuit cable and connect the live, neutral and earth conductors to their respective terminals **(3)**.

Prepare the end of the cable running from the light fitting and connect its brown conductor to the switch terminal **(3)**. Connect its blue conductor to the neutral terminal, and its sheathed earth conductor to the earth terminal.

Prepare the end of the cable running from the switch and then connect its brown conductor to the live terminal, the sleeved earth conductor to the earth terminal, and its blue conductor to the switch terminal **(3)**. Identify this last conductor by wrapping a piece of brown tape round it. Make sure all the connections are secure, and then refit the cover of the junction box.

Finally, put the lighting-circuit fuse back in the consumer unit and switch the power supply on again.

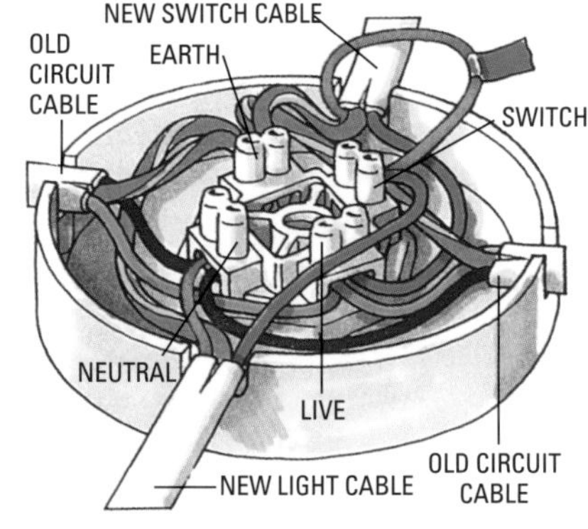

3 Wiring the junction box

• Old colour coding
When working on a house built before 2005 you are likely to find that the existing cables are colour-coded black for neutral and red for live. The diagram (left) shows new-style cable being connected to existing old-style circuit cable.

SEE ALSO: Building Regulations 8, Switching off 16, Running cable 23–5, Lighting circuits 47, Circuit lengths 66

Passive infra-red lighting

Exterior lighting connected to a passive infra-red detector illuminates only when the sensor picks up the body heat of someone within range. Because the detector is also fitted with a photocell, the lighting only operates at night.

A passive infra-red system has two advantages over simple dusk-to-dawn lighting. A porch light, for example, switches on only as you or visitors approach the entrance, then switches off again after a set period. Consequently, you are not wasting electricity by burning a lamp continuously all night. Secondly, remote passive infra-red detectors can be positioned to detect movement of an intruder almost anywhere around your home and will switch on all your security lights or only those you think necessary. The effect is likely to startle intruders and hopefully deter them from approaching any further.

Light fittings and sensors

Because infra-red detectors are designed to be mounted about 2.5m (8ft) above ground level, light fittings made with integral detectors tend to be for porch lighting and most of them are styled accordingly. However, since the lighting needs to be operated for periods of no more than a few minutes at a time, remote infra-red detectors are often used to control powerful halogen floodlights. Floodlights are also available with built-in detectors, which simplifies the wiring.

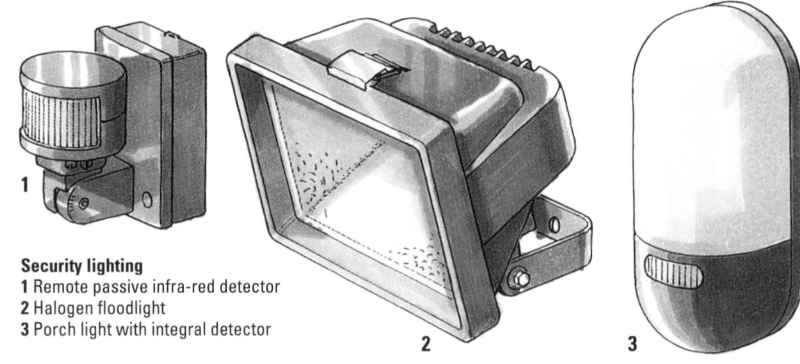

Security lighting
1 Remote passive infra-red detector
2 Halogen floodlight
3 Porch light with integral detector

Positioning detectors

Unless you position detectors carefully, your security lighting will be activated unnecessarily. This can be a nuisance to neighbours and, as with any security device that's constantly giving false alarms, you will soon begin to mistrust it and ignore its warnings.

Infra-red detectors have sensitivity controls so that they won't be activated by moving foliage or the presence of small animals. However, if your house is close to a footpath, you will need to adjust the angle of the detectors so that the lights don't switch on every time a pedestrian passes by.

When fitting a remote detector, make sure it isn't aimed directly at a floodlight that it is controlling – or its photocell will try to switch off the light as soon as it illuminates, and the likely result will be a light that simply flickers and never fully illuminates the scene.

It is also important not to position an infra-red detector above a balanced flue from a boiler, or any other source of heat that could activate the sensor.

The circuit

It is usually possible to wire infra-red security lighting in exactly the same way as dusk-to-dawn lighting. However, as some detectors are capable of controlling several powerful floodlights, you may not be able to run them from the domestic lighting circuit. In which case, you will need to run a 2.5mm^2 two-core-and-earth spur from a power circuit and control the lighting with a switched fused connection unit. Use a 3amp fuse in the connection unit for a combined rating of up to 690W; a 13amp fuse for anything greater.

Wiring a remote sensor

Unless the manufacturer's instructions suggest an alternative method, wire an individual light fitting with an integral infra-red detector in the same manner as a dusk-to-dawn fitting.

To wire a remote sensor controlling light fittings mounted elsewhere, take the incoming cable from the junction box or fused connection unit into the back of the fitting and connect its brown and blue conductors to the 'Mains' terminals **(1)**. Run a second cable of the same size from the 'Load' terminals **(1)** back through the wall and on to the first light fitting. Sleeve both bare copper earth conductors and connect them to the earth terminal **(1)**.

Wire the first light fitting using the method described for a dusk-to-dawn fitting, then connect another cable to the same terminals and run it on to the second light fitting, and so on.

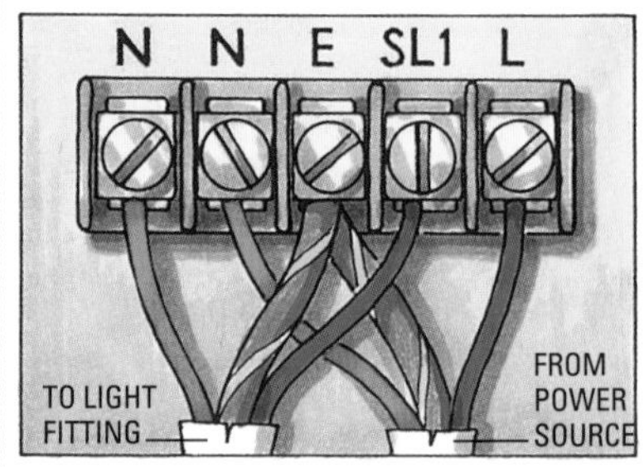

1 Wiring a remote sensor

Adjusting detectors

Having made all the connections, you need to set the infra-red detector's adjustment knobs or screws. One of them is for setting the photocell so that the system only operates during the hours of darkness. A second control dictates the period of time that the lights will remain on – three or four minutes should be sufficient to make any intruder feel conspicuous. Some sensors are fitted with a sensitivity control to avoid 'nuisance' operation.

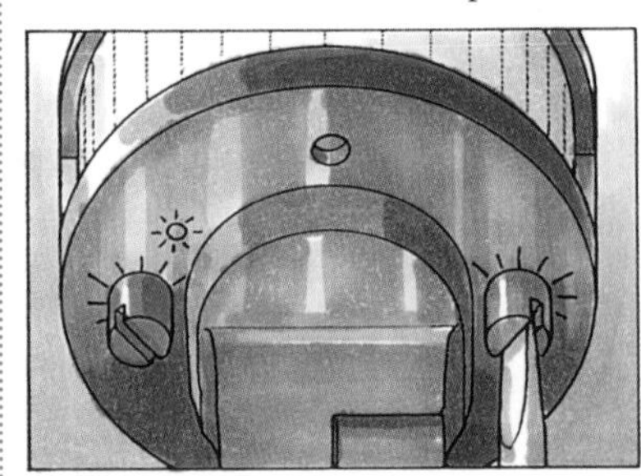

The final operation is to set the sensor's controls

SEE ALSO: Building Regulations 8, Switching off 16, Running cable 23–5, Running a spur 31, Fused connection units 34, Lighting circuits 47, Dusk-to-dawn lighting 59, Circuit lengths 66

Closed-circuit television

Surveillance by closed-circuit television acts as a deterrent to would-be intruders; and if you choose to record the output from your cameras, in the event of a burglary it may help the police to apprehend the perpetrator and recover your stolen property.

There are numerous CCTV systems at your disposal, including highly sophisticated equipment primarily designed for protecting commercial premises. However, if you want to install a system yourself, it may be best to choose one of the relatively inexpensive but effective DIY kits produced for the purpose. This type of kit includes all the accessories and materials you need, including a choice of black-and-white or colour cameras.

DIY kits

Not every CCTV kit contains the same equipment, but illustrated below is a selection of accessories sold for DIY installation. Light fittings are rarely, if ever, included in these kits, but since a great many attempted burglaries occur after dark, it's hardly worth the expense of installing CCTV unless you are prepared to buy and install compatible security lighting. A monitor is not required, as DIY kits are designed to be connected to ordinary television sets and video-cassette recorders.

Camera and cable

Power-supply unit

Scart connector

Switcher unit

Distribution box

Light and buzzer

Switcher unit remote control

Extension cable

VCR controller

Cameras
Up to four cameras can be connected to the average CCTV system. Black-and-white cameras are the least expensive and tend to give a sharper image in low light levels. Most cameras incorporate a microphone, and some have built in PIR (passive infra-red) movement detectors.

Power-supply unit
CCTV power-supply units have built-in 13amp plugs. There are heater elements in the cameras to prevent condensation, so they have to be connected to the electrical supply permanently. Power consumption, however, is negligible.

Distribution box
This compact unit sends the signals from the camera to your television set. You can make the connection using phono plugs, but a Scart connector is preferable.

Switcher unit
You will only need a switcher unit if you have more than one camera. Connected to the distribution box, it has a socket for each camera input.

VCR controller
This device switches on your VCR when a camera detects movement in the vicinity of your house.

Audible warning
A buzzer will alert you when a camera picks up a possible intruder, even when your TV set is switched off. The unit, which also includes a small flashing light, plugs into the distribution box.

Cable
Special colour-coded multi-core cable transmits the signals from the camera to your TV set. It is supplied in standard lengths, and extension cables are available.

SEE ALSO: Building Regulations 8, Security lighting 59–60, Installing CCTV 62

Installing CCTV

Before you go to the trouble of making a permanent installation, connect all the components together to ensure everything works and that they transmit a clear image to your television set.

Positioning cameras

Try to cover the most likely approaches to your home. Adjust each camera so that it is aimed at a slight angle to the route an intruder might take. It will then record him from several different angles as he passes by. This may help the police identify a known burglar.

Place your CCTV cameras out of reach – somewhere between 2.5 and 3m (8 to 10ft) from the ground. Point each camera down at an angle, never directly into the sun or towards a light fitting. If you have PIR detectors fitted to the cameras, make sure they are adjusted to avoid false alarms being triggered by passing cats and foxes and moving branches.

If you're in doubt about the suitability of a particular location, it may make sense to rig up a temporary connection and test the camera before you install permanent wiring.

Security lighting
To use cameras effectively after dark, install good porch lighting or, better still, floodlights fitted with PIR detectors. Since some cameras are particularly sensitive to infra-red, it pays to choose fittings that take halogen or tungsten bulbs rather than fluorescents.

Try to achieve even illumination – it is difficult for a security camera to cope with strong contrasts between, say, a dark carport and a well-lit pathway.

Running and connecting cable

Run the cable supplied with the kit from the camera to the television set, keeping as much of the wiring indoors as possible. Clip the cable to a sound surface at 1m (3ft) intervals, making sure it does not run alongside mains power cables.

To make it more difficult for anyone to tamper with the connections, feed the cable through a hole drilled directly behind the camera's housing.

It is inadvisable to coil up excess cable. Instead, cut it to length and feed the cut end though the grommet or seal in the camera mounting; then prepare and connect the colour-coded wires to the camera terminal box, following the manufacturer's instructions.

Installing the distribution box

Install the distribution box behind your television set. Plug the Scart connector into the back of the set, and the DIN connector on the camera cable into the distribution box. Then connect the power-supply unit to the distribution box and plug it into a convenient 13amp socket.

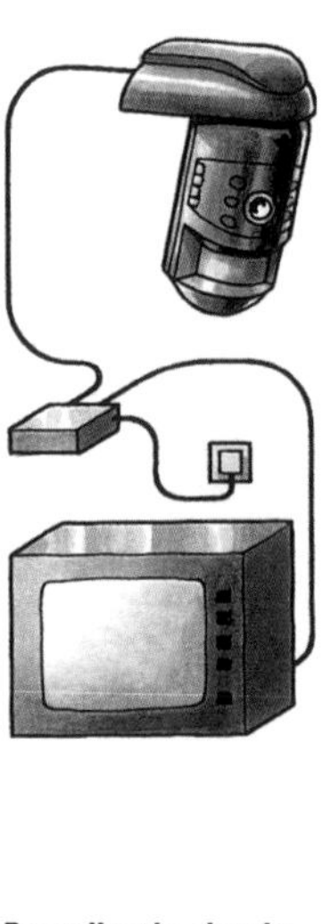

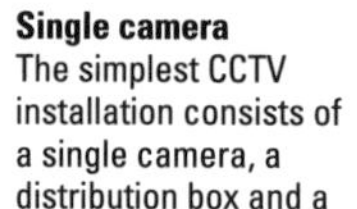

Single camera
The simplest CCTV installation consists of a single camera, a distribution box and a television set.

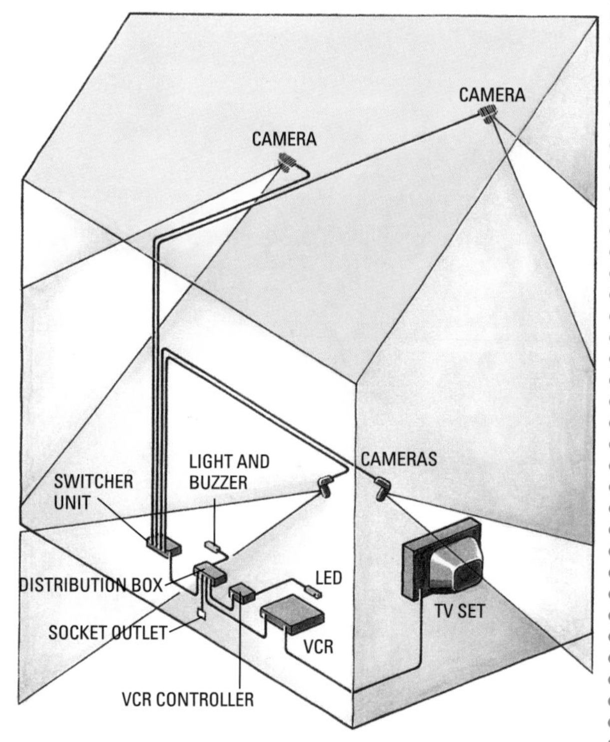

Recording the signals
With a more complex installation, you can have several cameras connected to a VCR so that you can make a taped record of would-be intruders.

RECORDING INTRUDERS

If you want to tape the pictures transmitted from your cameras, plug the Scart connector into a video recorder instead of the TV set. Switch the TV to the video channel, and the VCR to the AUX (auxiliary) channel.This will give you the option to keep a record of anyone who approaches your house, even when you are not at home.

Switching channels

Provided that you leave your TV set switched to the video channel, your TV viewing will be interrupted to show you the scene outside as soon as a camera detects an intruder. You will still be able to switch from one TV channel to another, using the standard VCR remote controller.

Recording the scene

If you want the VCR to start recording whenever the camera's PIR detects movement, fit a special VCR controller to the distribution box and point its infra-red output at the port used by your standard VCR remote controller.

If you prefer, you can fit a small unobtrusive LED (light-emitting diode) extension to the controller, so that you can hide the main unit out of sight.

You can preset how long you want the VCR to continue recording before it automatically switches off.

Using a second VCR

With your VCR switched to the AUX channel, you won't be able to view a prerecorded tape – so you may want to buy a cheap second-hand VCR that you can use solely for surveillance. You could install this VCR somewhere out of sight, so that a burglar is less likely to spot it and destroy the evidence.

Linking to several TV sets

You can use all the television sets in your home as surveillance monitors, but to do this the signals have to be transmitted via the TV aerial. For multi-set monitoring, you need to connect the distribution box to a device known as a modulator. Connect the TV aerial to the modulator, and the modulator to your existing aerial splitter socket.

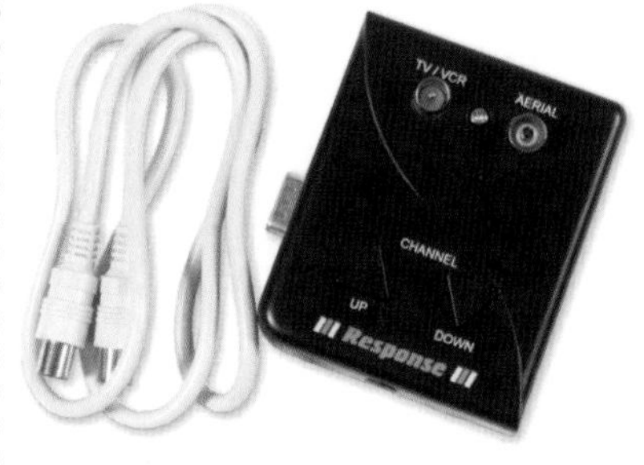

A modulator allows you to monitor several TV sets

SEE ALSO: Building Regulations 8, Wiring TV-aerial sockets 43, Security lighting 59–60

Garden lighting and pumps

Just a few outdoor lights can transform a garden dramatically. Spotlights or floodlighting can be used to emphasize particularly attractive features, at the same time providing functional lighting for pathways and steps, while strings of light bulbs threaded through foliage afford attractive background illumination.

Some of the most impressive effects are produced with underwater lights, which can be used to make small pools or fountains the focal points of a garden.

Extra-low-voltage lighting

Some types of garden light fitting can be powered directly from mains electricity, though they need to be installed by a professional electrician. However, you can install light fittings, or a complete lighting kit, yourself if they connect up to an extra-low-voltage transformer.

Position the transformer indoors near to a 13amp socket outlet, in a garage or workshop, and connect it to the socket by an ordinary square-pin plug. The flex – which is normally supplied with the light fitting – is connected to the two 12V outlet terminals on the transformer. Carry out the connections to the lights in accordance with the manufacturer's instructions.

Unless the maker states otherwise, extra-low-voltage flex supplying garden lights can be run along the ground without further protection – but inspect it regularly and don't let it trail over stone steps or other sharp edges likely to damage the PVC insulation if someone steps on it. If you have to add extra flex, use a waterproof connector.

Pool lighting

Pool lights are normally submerged so as to have at least 18mm (¾in) of water above the lens. Some are designed to float unless they're held down below the surface by smooth stones, carefully placed on the flex.

Submerged lights get covered by the particles of debris that float in all ponds. To clean the lenses without removing the lights from the water, simply direct a gentle hose over them.

You will find that occasionally you have to remove a light and wash the lens thoroughly in warm soapy water. Always disconnect the power supply before you handle the lights or take them out of the pond.

Run the flex for pool lighting under the edging stones via a drain made from corrugated plastic sheeting. The entire length of the flex can be protected from adverse weather by being run through a length of ordinary garden hose. Take the safest route to the power supply, anchoring the flex gently in convenient spots – but don't cover it with grass or soil in a place where someone might inadvertently cut through the flex with a spade or fork. Join lengths of low-voltage cable with waterproof connectors.

Pumps

Electric pumps can be used in garden pools to create fountains and waterfalls. A combination unit will send an adjustable jet of water up into the air and at the same time pump water through a plastic tube to the top of a rockery to trickle back into the pool.

Some pumps run directly from the mains supply. To fit these, follow the manufacturer's instructions and consult an electrician. But there are also extra-low-voltage pumps that connect to a transformer (see left) shielded from the weather. So you can disconnect the pump without disturbing the extra-low-voltage wiring to the transformer, join two lengths of cable with a waterproof connector. Conceal the connector under a stone or gravel beside the pool.

Most manufacturers recommend you take a pump from the water at the end of each season and clean it thoroughly, then return it to the water immediately. To avoid corrosion, don't leave it out of the water for very long without cleaning and drying it. Never service a pump without first disconnecting it from the power supply. During the winter, run the pump for at least an hour every week, to keep it in good working order.

- **Connected kits**
You do not have to notify your Building Control Officer if you are installing a complete ready-connected lighting kit or pump that is CE approved and that is not part of the fixed wiring of the house.

- **'Extra-low-voltage'**
Strictly speaking, this is the correct term to describe equipment that runs on 50V or less. However, manufacturers and suppliers often use the term low-voltage to describe equipment of this kind.

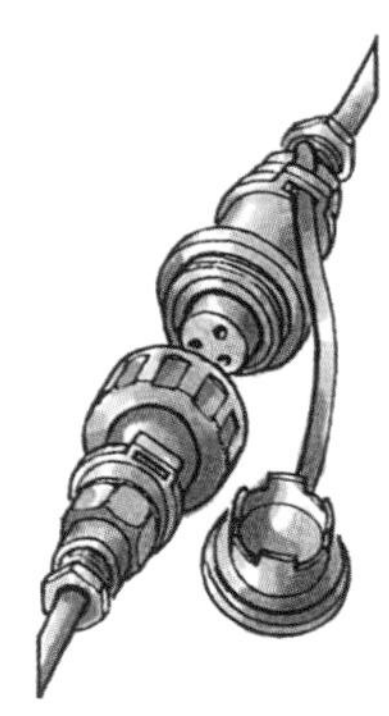

Waterproof cable connector
You can obtain suitable cable connectors from pump and lighting suppliers.

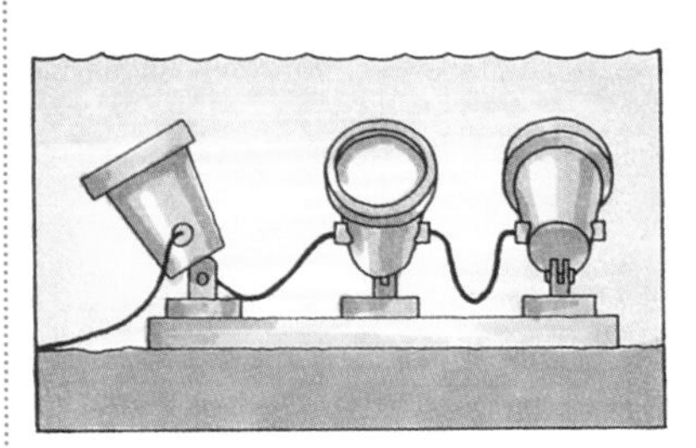

Stand underwater floodlights on a flat stone

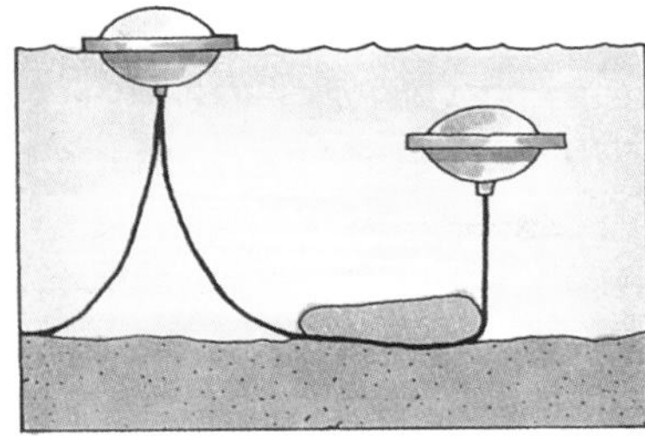

To submerge a light, place a stone on the cable

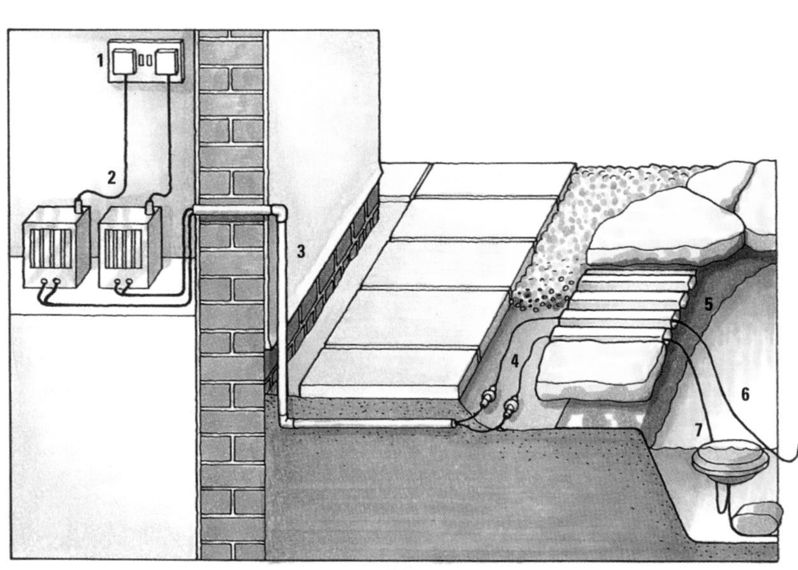

Pump and lighting circuits
1 Socket outlet
2 Isolating transformers BS 3535 Type 3 (also numbered BS EN 60742).
3 Plastic conduit
4 Waterproof connectors
5 Home-made drain
6 Pump cable
7 Lighting cable

SEE ALSO: Building Regulations 8

Running power to outbuildings

The power supply to a separate workshop, garage or tool shed can't be tapped from other domestic circuits. The cable has to run either from a switchfuse unit or from its own fuseway in the consumer unit and pass safely underground or overhead to the outside location – where it must be wired into a switchfuse unit from which the various circuits in the outbuilding can be distributed as required.

1 2 3

Outdoor cables
1 Armoured cable
2 Mineral-insulated copper-sheathed cable
3 PVC-insulated-and-sheathed cable

Types of cable permitted outdoors

Three types of cable can be used outside. The type you choose will depend on how you wish to run the cable.

Armoured cable

Although insulated in the ordinary way, in addition this two-core or three-core cable is protected by steel-wire armour that is insulated with an outer sheath of PVC. With two-core cable, the metal armour provides the path to earth, but as some authorities insist on two-core-and-earth cable, check what is required before you buy your cable.

Armoured cable is expensive and has to be terminated at a special junction box at each end of its run, where it can be connected to ordinary PVC-insulated cable. It is fitted with threaded glands for attaching it to the junction boxes. When buried in the ground, this type of cable must be covered with cable covers or warning tape.

Mineral-insulated copper-sheathed cable

The bare copper conductors of mineral-insulated copper-sheathed (MICS) cable are tightly packed in magnesium-oxide powder within a copper sheathing. The copper sheathing can act as the earth conductor and is itself sheathed in PVC insulation. Because the mineral powder absorbs moisture, special seals must be fitted at the ends of the cable.

Like armoured cable, MICS cable is expensive and has to be terminated at special junction boxes so that cheaper cable can be used in the outbuilding itself. It must also be protected with cable covers or warning tape when it is buried below ground.

PVC-insulated-and-sheathed cable

Ordinary PVC-insulated two-core-and-earth cable can be run underground to an outbuilding – but only if it is protected with impact-resistant heavy-gauge conduit and paving slabs that will, together, provide at least the same degree of mechanical protection as armoured cable.

If the conduit has to go round corners, elbow joints can be cemented onto the ends of straight sections. The cable itself should be continuous.

PVC-insulated cable can also be run overhead quite safely – but only under certain specified conditions (see below).

Ways of running outdoor cable

Underground

Running cable underground is usually the best way of supplying electricity to an outbuilding.

You should bury the cable in a trench at least 500mm (1ft 8in) deep, or deeper still if the cable has to pass underneath a vegetable plot or flowerbed, or other areas where digging is likely to go on.

It's best to plan your cable run so as to avoid such areas wherever possible. But you can provide extra protection for the cable by laying housebricks along both sides of it, supporting a covering made from pieces of paving slab. You should also bury special black-and-yellow-striped tape to serve as a warning to anyone who happens to uncover the slabs at a later date.

Line the bottom of the trench with finely sifted soil or sand, lay the cable or conduit, and then carefully fill in.

Protecting underground cable
Support paving slabs on bricks to protect a cable or conduit at the bottom of a trench. Lay marking tape on top of the slabs. Lay another strip of tape just below ground level to warn anyone who may be digging in this area in the future.

Overhead

Ordinary PVC-insulated cable can be run from house to outbuilding provided that it is at least 3.5m (12ft) above the ground or 5.2m (17ft) above a driveway that's accessible to vehicles. The cable may not be used unsupported over a distance of more than 3m (10ft), though the same distance can be spanned by running the cable through a continuous length of rigid steel conduit suspended at a height of at least 3m (10ft) above the ground or 5.2m (17ft) above a driveway. The conduit itself must be earthed.

Over greater distances, the cable must be supported by a metal catenary wire stretched taut between the house and outbuilding. The supporting wire must be earthed. The cable is either clipped to it or hung from slings. PVC-insulated cable can also be run through conduit mounted on a wall.

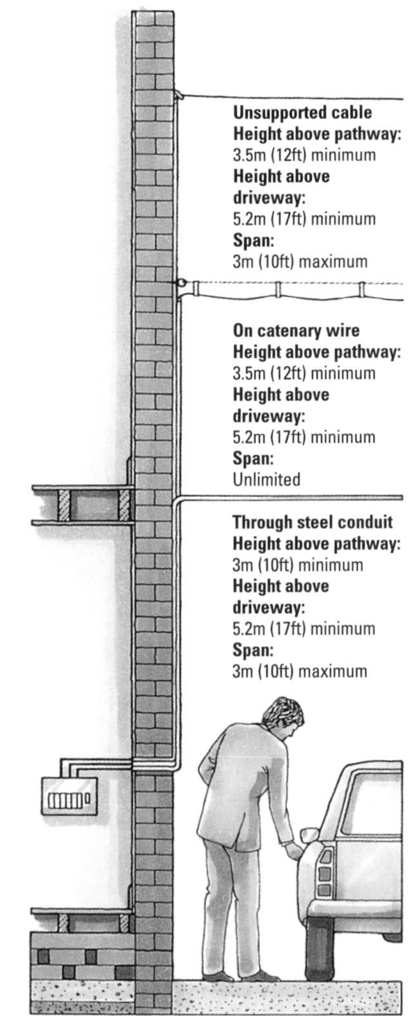

Running cable overhead

SEE ALSO: PVC-insulated cable 22

Running the circuit N

A variety of equipment and cables can be used to run a circuit to an outbuilding. The method described here uses normal PVC-insulated cable and a switchfuse unit at each end of the circuit – but the cable can run from a spare fuseway in the consumer unit if one is available.

It's assumed here that both sockets and lighting are required in the outbuilding, so the lighting circuit is taken from the power cable via a junction box and an unswitched fused connection unit. Run the cable underground in impact-resistant heavy-gauge plastic conduit protected by paving slabs, ensuring that it enters both buildings above the DPC and, if possible, beneath the floorboards.

House end of the circuit

Mount a 30amp switchfuse unit near the meter and then fit a 30amp circuit fuse. Install a residual current device between the unit and the meter.

Next, run 10mm^2 two-core-and-earth cable from the 'Load' terminals of the RCD to the 'Mains' terminals of the switchfuse unit. Connect the outgoing 4mm^2 cable to the 'Load' terminals **(1)** of the switchfuse unit.

Prepare one brown and one blue 16mm^2 PVC-sheathed-and-insulated cable to serve as the meter leads and attach them to the 'Mains' terminals of the RCD.

Wire a 16mm^2 earth lead to the RCD **(1)** in readiness for connection to the consumer's earth terminal. But don't try to make the connections to the meter or the electricity company's earth yourself – these must be made by the company.

Outbuilding end of circuit

Run a 4mm^2 two-core-and-earth cable through conduit from the house to the outbuilding, terminating at a 30amp switchfuse unit mounted on the wall.

Connect up the incoming cable to the supply or 'Mains' terminals of the switchfuse unit and the outgoing 4mm^2 cable to its 'Load' terminals **(2)**, then run this cable to the outbuilding's sockets.

Insert a 30amp junction box at some point along the power cable **(3)**, and run a 4mm^2 spur from it to an unswitched fused connection unit fitted with a 3amp fuse. Then run a 1mm^2 two-core-and-earth cable from the connection unit to the light fitting and switch.

Workshop circuit
1 Meter
2 RCD
3 Switchfuse unit
4 4mm^2 cable
5 Conduit
6 Switchfuse unit
7 Junction box
8 Socket outlet
9 Fused connection unit
10 Lighting junction box
11 Light fitting
12 Light switch

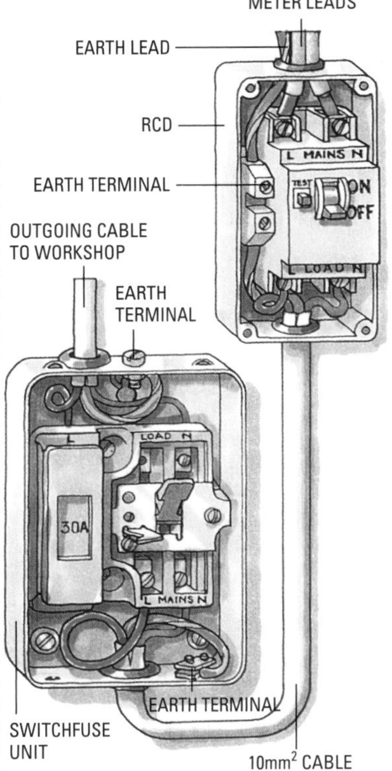

1 Wiring switchfuse unit and RCD

• Meter leads and earth lead
If 16mm^2 cable is too thick for the terminals in the RCD, use 10mm^2 cable – but keep the leads as short as possible.

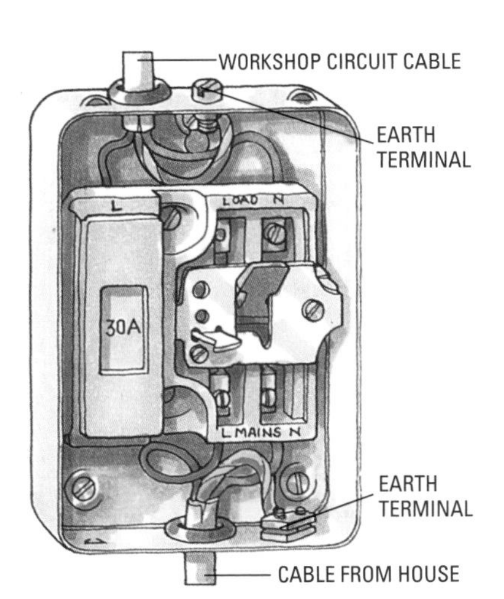

2 Wiring workshop switchfuse unit

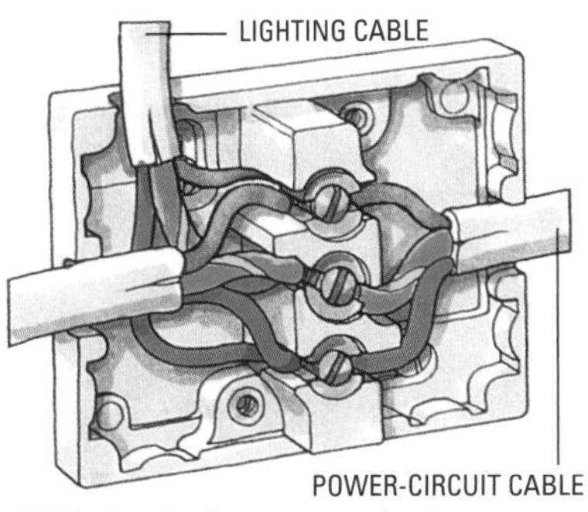

3 Wiring junction box on power circuit

SEE ALSO: Building Regulations 8, Running cable 23–5, Connecting sockets 30, Fused connection units 34, Lighting circuits 47, Lighting junction box 48, One-way switch 52

Complete wiring

● **BCO inspections**
Hiring a professional electrician will save you the trouble of having to submit your plans for rewiring – and will also save you the cost of inspection by the BCO.

Planning ahead

Before deciding to take on the complete rewiring of your house yourself, you need to consider the time factor carefully. When you are working on a single circuit, the rest of the household can function normally, but to renew all the circuits means that every part of your home will, eventually, be affected.

A full-time professional can cope with all this in such a way that the level of inconvenience is kept to a minimum. But an amateur will almost certainly have to think in terms of a time span lasting several weeks – especially since it is very important not to work hastily on such installations. Hurried work can lead to dangerous mistakes!

So unless you are very experienced and are able to make the installation a full-time commitment for at least a week or two, you would be well advised to employ a fully qualified electrician to undertake this time-consuming job.

He or she may perhaps allow you to work alongside – which could mean a considerable saving on the cost if you are able to carry out some of the jobs that have nothing to do with electricity, such as running cable under floors and channelling out plaster and brickwork.

Circuits: maximum lengths

The maximum length of a circuit is limited by the permitted voltage drop and the time it takes to operate the fuse or MCB in the event of an earth fault.

The method for calculation given in the Wiring Regulations is extremely complicated – but the table below will provide you with a simple method for determining the maximum cable lengths for common domestic circuits.

If necessary, split up your circuits so that none of the indicated cable lengths are exceeded. If your requirements fall outside the limits of this chart, then ask a professional electrician to make the calculations for you.

Rewirable fuses are not included, as they are subject to special restrictions – which makes them an unwise choice.

Most two-core-and-earth cables have a standard-size protective circuit conductor (earth wire). In each case, the chart shows the size of earth wire used in the calculations.

The maximum circuit lengths given in the chart are based on the assumption that you won't install any cables where the ambient temperature exceeds 30°C (86°F), that no cables will be bunched together, and that you will not cover any of the cables with thermal insulation.

The shower-circuit lengths assume that a 30 milliamp RCD is used in the circuit. The cooker-circuit lengths allow for using a control unit with a built-in socket outlet. However, you can safely use the same figures for wiring a control unit without a built-in socket.

MAXIMUM LENGTHS FOR DOMESTIC CIRCUITS

TYPE OF CIRCUIT	Max. floor area in sq m	Cable size in mm²	Size of earth wire in mm²	Current rating of circuit fuse (USING FUSES)	Max. cable length using cartridge fuse (USING FUSES)	Current rating of MCB (USING MCBs)	Max. cable length using MCB (USING MCBs)
RING CIRCUIT	100	2.5	1.5	30amp	68m	32amp	68m
RADIAL CIRCUIT	20	2.5	1.5	20amp	37m	20amp	34m
	50	4	1.5	30amp	19m	32amp	21m
COOKER up to 13.5kW		4	1.5	30amp	19m	32amp	21m
COOKER from 13.5 to 18kW		6	2.5			40amp	27m
IMMERSION HEATER up to 3kW		2.5	1.5	15amp	39m	16amp	39m
SHOWER up to 10.3kW		10	4	45amp	46m	45amp	46m
SHOWER from 10.3 to 10.8kW		10	4			50amp	44m
STORAGE HEATER up to 3.375kW		2.5	1.5	15amp	34m	16amp	34m
STORAGE FAN HEATER up to 6kW		4	1.5	30amp	32m	32amp	32m
FIXED LIGHTING excluding switch drops		1	1	5amp	83m	6amp	83m
		1.5	1	5amp	126m	6amp	126m

Before discussing your requirements with a professional electrician, you need to form clear ideas about the kind of installation you want. Although you may eventually decide between you to change some of the details, a proper specification can be of considerable help to the electrician – and should also enable you to avoid expensive additions and modifications.

Choosing the best consumer unit

It is worth installing the best consumer unit you can afford. Choose one that has cartridge fuses or miniature circuit breakers (MCBs), and make sure it has enough spare fuseways for possible additional circuits.

Residual current devices

Ask the electrician about the possibility of installing a residual current device (RCD). You could have one built into your consumer unit.

Power circuits

Ring circuits are better than radial circuits for supplying socket outlets. Provided that the floor area in question does not exceed 100sq m (120sq yds), you can have as many sockets as you want – so make sure your plan includes enough outlets to meet your present and likely future needs. Economizing on the cost of a few sockets now could cause you considerable inconvenience in the future, if you have to start adding spurs to the system.

Lighting circuits

Modern domestic lighting circuits are normally designed around a loop-in system; but you can supply individual light fittings from junction boxes if that is the most practical solution.

You should insist on a lighting circuit for each floor – so that you will never be left totally without electric lights if a fuse should blow.

In the interests of safety, make sure you have two-way or three-way switches installed for lights in passageways and on landings and staircases.

Additional circuits

If you are having your whole house rewired, consider installing extra radial circuits for appliances such as immersion heaters and electrically heated showers.

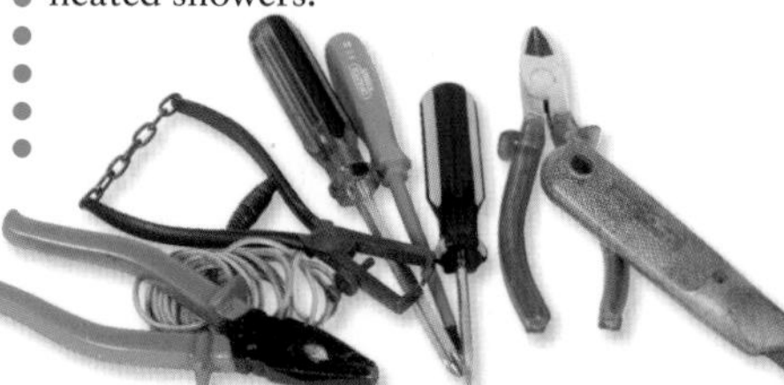

SEE ALSO: Building Regulations 8, RCDs 17, Consumer units and fuses 18–9, Lighting circuits 21, Power circuits 21, Fixed appliances 34–5, Wiring a cooker 37–8

Electrical tools

ELECTRICIAN'S TOOL KIT

You need only a fairly limited range of tools to make electrical connections, but an extensive general-purpose tool kit is required for making cable runs and for fixing electrical accessories and appliances to the structure of the house.

SCREWDRIVERS

Buy good screwdrivers for tightening electrical terminals. Cheap ones are practically useless, being made from such soft metal that the tips soon twist out of shape.

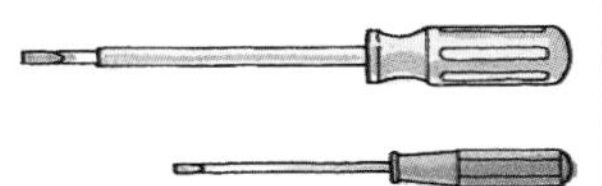

Terminal screwdriver

A terminal screwdriver has a long, slim cylindrical shaft that is ground to a flat tip.

For turning screw terminals in sockets and larger appliances, buy a screwdriver with a plastic handle and a plastic insulating sleeve on its shaft.

Use a smaller screwdriver with a very slim shaft to work on ceiling roses or to tighten plastic terminal blocks in small fittings.

Cabinet screwdriver

You will need a woodworking screwdriver to fix mounting boxes to walls.

SKATE

An electrician's skate has a cutting disc that severs the joint between tongue-and-groove floorboards. Run the tool back and forth with one foot.

WIRE CUTTERS

Use wire cutters for cropping cable and flex to length.

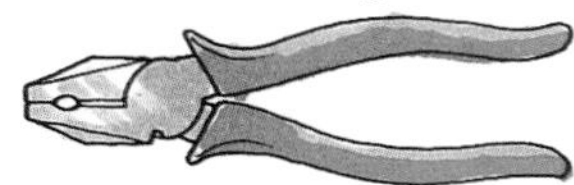

Electrician's pliers

These are engineer's pliers with insulating sleeves shrunk onto their handles. You can use pliers to crop circuit conductors.

Diagonal cutters

Diagonal cutters will crop thick conductors more effectively than electrician's pliers, but you may need a junior hacksaw to cut meter leads.

WIRE STRIPPERS

There are various tools for cutting or stripping the plastic insulation that covers cables and flexible cords.

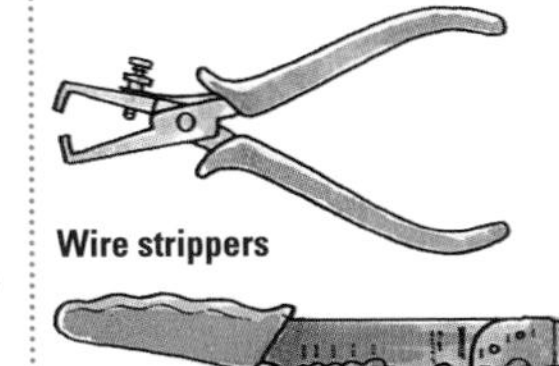

Wire strippers

Multi-purpose tool

Wire strippers

To remove the insulation from cable and flex, use a pair of wire strippers with jaws shaped to cut through the covering without damaging the wire core. There is a multi-purpose version that can both strip the insulation and crop conductors to length.

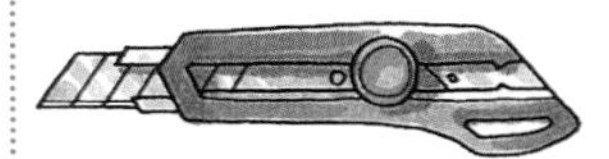

Sharp knife

A knife with sharp disposable blades is best for slitting and peeling the sheathing encasing cable and flex.

DRILLS

When you run circuit wiring, you need a drill with several special-purpose bits for boring through wood and masonry.

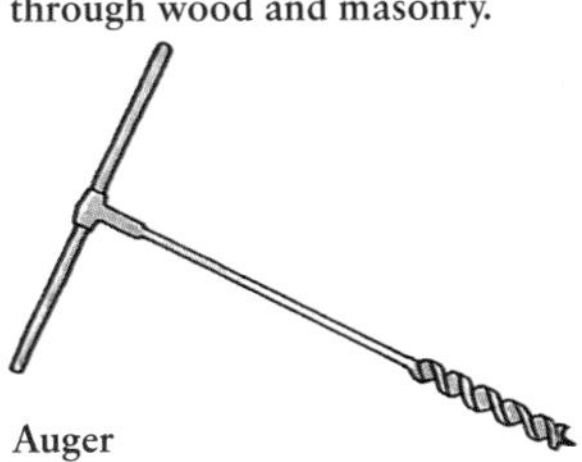

Auger

Some electricians employ a long wood-boring auger to drill through the wall head plate and noggings when they're running a switch cable from an attic down to its mounting box.

Power drill

A cordless power drill is ideal for boring cable holes through timbers and for making wall-plug fixings. As well as standard masonry bits for wall fixings, you will need a much longer version for boring through brick walls and clearing access channels behind skirting boards.

If you shorten the shaft of a wide-tipped spade bit, you can use it in a power drill between floor joists, instead of hiring a special joist brace.

TESTERS

Even when you have turned off the power at the consumer unit, use a tester to check that the circuit is safe to work on.

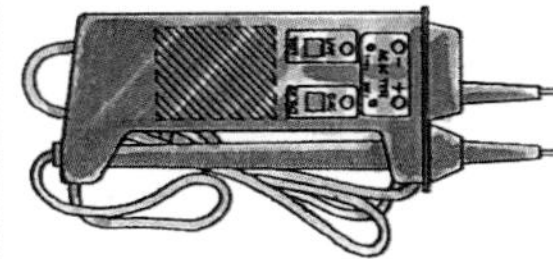

Electronic mains tester

Be sure to buy a two-prong electronic tester that's intended for use with mains voltage – similar devices are sold in auto shops for 12V car wiring only.

Always check that the tester is functioning properly before and after you use it, by testing it on a circuit you know to be live.

Following the manufacturer's instructions, place one probe on the neutral terminal and the other one on the live terminal that is to be tested. If the bulb illuminates, the circuit is live; if it doesn't illuminate, try again between the earth terminal and each of the live and neutral terminals. If the bulb still doesn't light up, you can assume the circuit is not live, provided you have checked the tester.

Continuity tester

A continuity tester will test whether a circuit is complete or an appliance is properly earthed. Alternatively, buy a multi-tester that combines the functions of continuity testing and mains-voltage testing (see above).

Using a continuity tester

Switch off the power at the consumer unit before making the following test.

To find the two ends of a buried disconnected cable, twist the black (blue) and red (brown) conductors together at one end, and then apply the tester's probes to the same conductors at the other end ***(1)****. Depress the circuit-testing button on the tester. The bulb should light up and, with some testers, there may also be an audible signal. Untwist the conductors. Then make the test again – and if the bulb doesn't illuminate, the two ends belong to the same cable.*

To check whether a plug-in appliance is safely earthed, apply one probe to the earth pin of the plug – the longest of the three – and touch an unpainted part of the metal casing of the appliance with the other probe ***(2)****. Depress the test button – and if the earth connection is good, the bulb will illuminate. Don't try to use the appliance if the bulb illuminates when you apply the probe to either of the plug's other pins* ***(3)*** *– make sure the plug fuse is working. Have a suspect appliance overhauled by an electrician.*

You cannot test a double-insulated appliance, as it has no earth connection in the plug.

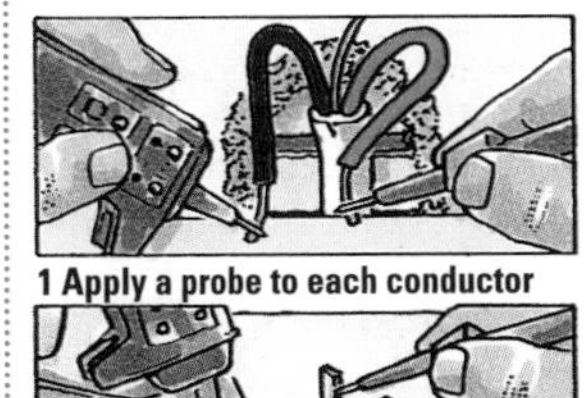

1 Apply a probe to each conductor

2 Test earth pin and casing

3 Test one other pin and casing

Final tests for fixed wiring

You can carry out simple tests yourself to check the effectiveness of the fixed wiring of your house, using equipment available from DIY outlets. However, compliance with the regulations requires final testing with prescribed specialist equipment to check readings that can be entered on certificates. Building Control inspectors and other professionals will use these instruments.

GENERAL-PURPOSE TOOLS

Every electrician needs tools for lifting and cutting floorboards, for fixing mounting boxes, and for cutting cable runs.

- Claw hammer
- Club hammer
- Cold chisel
- Bolster chisel
- Wood chisels
- Padsaw or power jigsaw
- Floorboard saw
- Spirit level
- Plasterer's trowel or filling knife (Either tool can be used for covering concealed cable with plaster or other kinds of filler.)
- Spanner (A small spanner is needed for making the earth connection in some appliances, and also for supplementary earth bonding.)

Torch
Keep a torch handy for checking your consumer unit when a fuse blows on a lighting circuit. You may also need artificial light when working on connections below floorboards or in the loft (a torch that stands unsupported is particularly helpful).

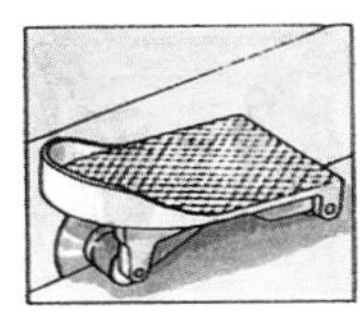

Electrician's skate

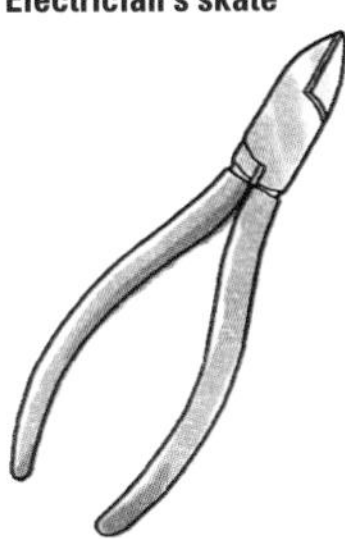

Diagonal cutters

● Essential tools
Terminal screwdrivers
Wire cutters
Wire strippers
Power drill and bits
Torch
Mains tester
Continuity tester
General-purpose tools

☛ SEE ALSO: Building Regulations 8, Double insulation 8, Using a tester 9, Supplementary bonding 10, Stripping flex 13, Stripping cable 22, Running cable 23–5, Using a skate 24, Drilling joists 25

Glossary

Accessory
An electrical component permanently connected to a circuit – a switch, socket outlet, connection unit etc.

Adaptor
A device that is used to connect more than one appliance to a socket outlet.

Ampere (Amp/A)
A unit of measurement of the flow of electric current necessary to produce the required wattage for an appliance.

Ceiling rose
A special junction box for connecting a suspended light fitting to a lighting circuit.

Ceiling switch
A light switch that is attached to a ceiling and operated by a pull-cord.

Chase
A groove cut in masonry or plaster to accept an electrical cable. *Or*
To cut such grooves.

Circuit
A complete path through which an electric current can flow.

Circuit breaker
A special switch installed in a consumer unit to protect an individual circuit. Should a fault occur, the circuit breaker will switch off automatically.

Conductor
A component, usually a length of wire, along which an electric current will flow.

Consumer unit
A box, situated near the meter, which contains the fuses of MCBs protecting all the circuits. It also houses the main isolating switch that cuts the power to the whole building.

Corrugated plastic sheet
A lightweight PVC sheet that is used to roof outbuildings and lean-to extensions.

Dimmer switch
A switch that changes the level of illumination by varying the electric current through a lamp.

Double-pole switch
A switch that breaks both the live and neutral conductors.

Downlighter
A type of ceiling-mounted light fitting that directs a relatively narrow beam of light to the floor.

Earth
A connection between an electrical circuit and the earth (ground).

Extension
A length of electrical flex for temporarily connecting the short permanent flex of an appliance to a wall socket.

Fuse
A protective device containing a thin wire that is designed to melt at a given temperature caused by an excess flow of current on a circuit.

Fuseboard
Where the main electrical service cable is connected to the house circuitry. The accumulation of consumer unit, meter etc.

Grommet
A ring of plastic or rubber lining a hole to protect an electrical cable from chafing.

Immersion heater
An electrical element designed to heat water in a storage cylinder.

Insulation – electrical
Nonconductive material surrounding electrical wires or connections to prevent the passage of electricity.

Insulation – thermal
Materials used to reduce the transmission of heat.

Miniature circuit breaker – MCB
See Circuit breaker.

Neutral
The section of an electrical circuit that carries the flow of current back to source. A terminal to which a connection is made.

Nogging
A short horizontal wooden member between studs in a timber-framed wall.

Protective multiple earth – PME
A system of electrical wiring in which the neutral part of the circuit is used to take earth-leakage current to earth.

Radial circuit
A power circuit feeding a number of socket outlets or fused connection units and terminating at the last accessory.

Residual current device – RCD
A device that monitors the flow of electrical current through the live and neutral wires of a circuit. When it detects an imbalance caused by earth leakage, it cuts off the supply of electricity as a safety precaution.

Ring circuit
A continuous power circuit starting at and returning to the consumer unit. Also known as ring main.

Rocker switch
A modern-style switch operated by a lever that pivots about its centre.

Sheathing
The outer layer of insulation surrounding an electrical cable or flex.

Short circuit
The accidental rerouting of electricity to earth that increases the flow of current and blows a fuse.

Spotlight
A light fitting that directs a narrow beam of illumination onto a specific object or area of wall or floor.

Spur
A short length of cable that feeds a socket outlet or fused connection unit by taking its power via another similar accessory.

Storage heater
A space-heating device that stores heat generated by cheap night-rate electricity, then releases it during the following day.

Studs
The vertical members of a timber-framed wall.

Supplementary bonding
The connecting to earth of exposed metal appliances and pipework within a bathroom or kitchen.

Terminal
A connection for an electrical conductor.

Transformer
A device that increases or decreases the voltage on an electrical circuit.

Uplighter
A light fitting that reflects illumination onto a ceiling.

Volt
A unit of measurement of 'pressure' that is provided by Electricity Company generators and which drives the current along the conductors.

Wiring Regulations
A code of professional practice laid down by the Institution

Index

Page numbers in *italics* refer to photographs and illustrations.

Index

Index